Caribou

Common shrew

Raccoon

Snowshoe hare

Virginia opossum

Tiger

National Geographic

BOOK OF
Mammals
Volume Two

Prepared by the
Special Publications
Division

National
Geographic Society,
Washington, D. C.

Contents

Mane not yet fully grown, a young male lion rests quietly with a female on a plain in Africa.

Volume One

Volume Two

PRECEDING PAGES: Long ivory tusks of a walrus glisten in the sun. Behind it, other walruses in Alaska swim in icy water and rest on the rocky shore. PAGE 305: River otter stands alert on the banks of a stream in Colorado. ENDPAPERS: Making tracks, mammals big and small leave their footprints. COVER: Bold and majestic, a tiger, largest of the cats, stares from a forest in India.

K

Kangaroo

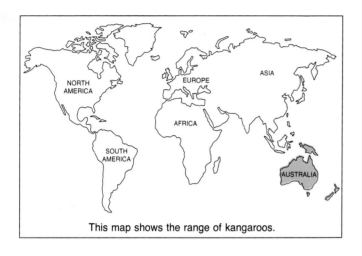

This map shows the range of kangaroos.

WHEN IT COMES TO CARRYING YOUNG, the kangaroo has an easy solution. It keeps its offspring in a pouch on its belly. Kangaroos are probably the best known of all pouched mammals, or marsupials (say mar-soo-pea-ulz). But few people realize that there are more than fifty kinds of kangaroos. These furry gray, brown, or reddish brown animals live in Australia, New Guinea, and on nearby islands. Some kinds live in New Zealand. Their ancestors were taken there from Australia in the 1800s.

Kangaroos come in many sizes. Two kinds, gray kangaroos and red kangaroos, may stand taller than most people and weigh more than 150 pounds

Red kangaroo: 57 in (145 cm) long; tail, 37 in (94 cm)

△ *Droopy-eyed and relaxed, a red kangaroo rests on a dry plain in Australia. More than fifty kinds of kangaroos live in Australia and on islands in the region.*

◁ *With a graceful takeoff, a red kangaroo springs into action. Using its powerful hind legs, it can reach heights of 6 feet (183 cm) and cover 25 feet (8 m) in a single jump.*

Two red kangaroos drink from a sheep rancher's pond. ▷ *In dry weather, these hardy animals get the moisture they need from the plants they eat.*

(68 kg). Another kind, the wallaroo (say WOLL-uh-roo), is slightly shorter and stockier. Red kangaroos, gray kangaroos, and wallaroos are known as the great kangaroos. They are the largest of all marsupials. Some other kinds of kangaroos seem tiny in comparison. The smallest kangaroo—the musky rat kangaroo—is only about the size of a rat.

Many kinds of small and medium-size kangaroos are called wallabies. Read about wallabies on page 554. Another member of the kangaroo family, the quokka, is described on page 470.

Kangaroos are found in many kinds of surroundings. Gray kangaroos live in forests. Red kangaroos roam plains. Wallaroos live in hilly, rocky regions. Most kinds of kangaroos feed only on grass. Some, though, nibble on other plants as well. The musky rat kangaroo eats insects and worms.

When grazing, a kangaroo moves slowly. First, it bends forward and balances on its front paws and tail. Then it swings its hind legs forward, following with its tail and front legs.

To cover short distances rapidly, a kangaroo hops. It pushes off with *(Continued on page 314)*

With joeys—young kangaroos—  tucked safely in their pouches, two gray kangaroo mothers look up from grazing. The adults use their muscular tails as props when they rest. The heads and feet of the joeys hang out because the young are almost too large for the pouches.

▽ To beat the heat, a young gray kangaroo licks its paws. As the moisture evaporates, it helps cool the animal.

Gray kangaroo: 48 in (122 cm) long; tail, 39 in (99 cm)

Kangaroo

◁ *Imitating its mother, a young joey hops to it! Bounding at a steady pace, the two gray kangaroos head for new feeding grounds. The kangaroos can keep up a speed of as much as 15 miles (24 km) an hour over long distances. About a year old, this youngster no longer returns to the pouch.*

▽ *Kangaroo mob scene: A mob—the name Australians give to groups of kangaroos—grazes in a forest clearing. The large adult male, second from right, probably leads this group of gray kangaroos. Red kangaroos also gather in mobs. Other kinds usually live alone or in pairs.*

Kangaroo

△ *Sheltered by ferns, vines, and other underbrush, a rufous rat kangaroo searches for food. At night, the tiny marsupial feeds on beetles and on woodland plants.*

Though often awkward on the ground, a tree kangaroo ▷ balances gracefully on a fallen limb. This tree dweller has long forelegs, unlike other kangaroos.

its strong hind legs and feet. As it sails through the air, its outstretched tail helps it balance. Larger kangaroos can travel 25 feet (8 m) in a single jump!

Females move very fast. Smaller and lighter than males, they can keep up a quick pace over a long distance. Female red kangaroos—called blue fliers because of their bluish fur—may travel more than 30 miles (48 km) an hour in short spurts.

A male kangaroo does not move quite so swiftly as a female does. With his broad, muscular chest and strong forelegs, however, he is better equipped to fight. Sometimes his enemy is the wild dog called the dingo. Rival kangaroos may fight for mates. But they rarely injure each other seriously. Some people think that a male kangaroo defends himself by boxing. He is really grabbing his rival with his forelegs. Rearing back on his tail, he kicks the foe with his

hind feet. Sometimes he bites his opponent's throat or uses his sharp claws to tear at the enemy.

Kangaroos mate at any time of year. Just a little more than a month after mating, a female bears one young. The tiny newborn, called a joey, is only about the size of a lima bean. Moments after its birth, the joey crawls into the safety of its mother's pouch. There it continues to grow and develop.

KANGAROO

LENGTH OF HEAD AND BODY: 10-65 in (25-165 cm); tail, 5-42 in (13-107 cm)

WEIGHT: 18 oz-150 lb (510 g-68 kg)

HABITAT AND RANGE: all kinds of habitats throughout Australia, New Guinea, New Zealand, and neighboring islands

FOOD: grasses, leaves, twigs, insects, and worms

LIFE SPAN: as long as 23 years in the wild

REPRODUCTION: 1 young after a pregnancy of about 1 month

ORDER: marsupials

After several months, the joey tumbles out of the pouch for the first time. If danger threatens, the young kangaroo dives headfirst back into the pouch. There a quick somersault turns it right side up. Red kangaroo young give up the pouch at about eight months of age. Grays remain for two or three months longer. Some rat kangaroos leave the pouch after only about three months.

Kangaroo rat

(say kang-guh-ROO RAT)

Desert kangaroo rat: 5 in (13 cm) long; tail, 7 in (18 cm)

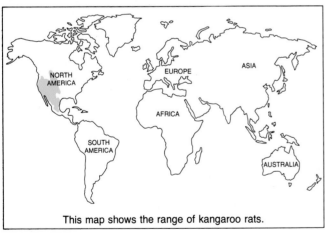

This map shows the range of kangaroo rats.

KANGAROO RAT

LENGTH OF HEAD AND BODY: 4-7 in (10-18 cm); tail, 5-9 in (13-23 cm)

WEIGHT: 1-6 oz (28-170 g)

HABITAT AND RANGE: deserts and dry, brushy regions in parts of North America

FOOD: seeds, leaves, stems, and insects

LIFE SPAN: as long as 9 years in captivity

REPRODUCTION: 1 to 6 young after a pregnancy of about 1 month

ORDER: rodents

Cheek pouches full, a desert kangaroo rat stands on a patch of grass in California. It uses the fur-lined pockets on the sides of its face to carry seeds to its burrow.

HOPPING ACROSS DESERT SAND, the kangaroo rat moves like a tiny kangaroo. A long, tufted tail helps the animal steer and keep its balance.

About twenty kinds of kangaroo rats live in parts of North America. Like many other desert rodents, kangaroo rats are well adapted, or suited, to their environment. Large hind feet keep them from sinking into the sand. The animals drink little or no water. They get moisture from the plants, insects, and seeds they eat.

When a kangaroo rat finds food, it may stuff some of it into its fur-lined cheek pouches. Then it carries the food to its burrow for storage.

Kangaroo rats spend most of the day underground. They live in burrows with several openings. These may serve as escape hatches when the animal is chased out by a fox, a coyote, or a snake.

A female kangaroo rat may have three litters a year. There are one to six young in each litter.

Kinkajou

△ *At home in the trees, a kinkajou moves swiftly along a narrow branch in a rain forest in Costa Rica. Because it can turn its hind feet backward, a kinkajou can scamper headfirst down a tree trunk with ease. The animal can hold on to a limb with its strong, slender tail.*

Nose buried in a hole, a young kinkajou sniffs for ▷ *insects. When eating fruit, the kinkajou uses its front paws and sharp claws to hold on to its meal.*

ITS TAIL TWINED AROUND A BRANCH, a golden-brown kinkajou hangs upside down in a tree. Like some kinds of monkeys, a kinkajou can hold on with its strong, slender tail. As the animal moves about the forests of Mexico and Central and South America, its tail serves as another paw, gripping a tree limb tightly. The 20-inch (51-cm) tail—about as long as the kinkajou's body—is useful in other ways. Running along a branch, a kinkajou swings its tail for balance. The animal wraps its tail snugly around its body when it sleeps.

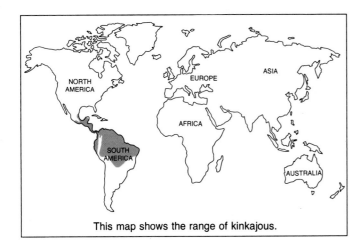
This map shows the range of kinkajous.

△ *Hungry kinkajou feasts on termites from a nest in Colombia. With its long, thin tongue, it scoops out the insects. The animal also uses its tongue to lick nectar from flowers and honey from bees' nests.*

Kinkajous are members of the raccoon family. Find out about raccoons on page 477. Like their raccoon relatives, kinkajous have nimble front paws. They can easily pluck fruit while hanging by their hind feet and tails. They clutch the food in one paw and break it into small pieces with the other.

As they feed, kinkajous often use their long, narrow tongues. The tongues can slide into bees' nests to lick out honey, one of the kinkajou's favorite foods. Kinkajous are even nicknamed honey bears because of their fondness for the sweet food. They also eat fruit and small animals.

The kinkajou is difficult to study in the wild. It spends most of its life among the branches. During hot tropical days, a kinkajou naps in a hollow tree. At night, it wakes up and scampers about. Usually it travels alone. But sometimes two or more kinkajous gather in a tree full of ripe fruit. Though the animals are rarely seen, people sometimes hear them growling and barking as they feed.

Kinkajous communicate by scent as well as by sound. Each animal has scent glands at the corners of its mouth and on its throat and belly. These produce a substance that the animals rub off on tree branches throughout the area where they live. Scientists think that kinkajous keep track of each other by means of these scent marks. The scents may also help the animals find mates.

A female kinkajou usually gives birth to one offspring. With its soft, tan fur and tightly shut eyes, the tiny animal looks like a newborn kitten. Its tail can already grip objects lightly, though it is not yet very strong. About one month later, the young kinkajou is able to see. It can hang head down by its tail at about three months of age. The animal is fully grown within a year.

KINKAJOU

LENGTH OF HEAD AND BODY: 17-22 in (43-56 cm); tail, 16-22 in (41-56 cm)

WEIGHT: 3-7 lb (1-3 kg)

HABITAT AND RANGE: forests in southern Mexico, Central America, and parts of South America

FOOD: honey, fruit, nectar, insects, birds, and small mammals

LIFE SPAN: as long as 19 years in captivity

REPRODUCTION: usually 1 young after a pregnancy of 3½ months

ORDER: carnivores

Klipspringer
The klipspringer is a kind of antelope. Read about antelopes on page 52.

Koala

△ *Nibbling on a leaf, a furry koala feeds in a eucalyptus tree in Australia. A strong-smelling oil in the leaves gives the animal an odor similar to cough drops.*

KOALA

LENGTH OF HEAD AND BODY: **24-33 in (61-84 cm)**

WEIGHT: **about 20 lb (9 kg)**

HABITAT AND RANGE: **forests of eastern Australia**

FOOD: **primarily eucalyptus leaves and shoots**

LIFE SPAN: **about 20 years in the wild**

REPRODUCTION: **usually 1 young after a pregnancy of about 1 month**

ORDER: **marsupials**

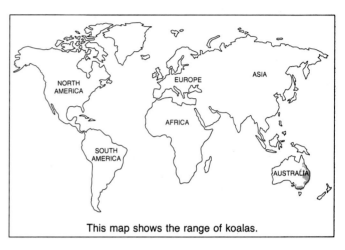

This map shows the range of koalas.

MOVING SLOWLY among the branches of a tree, the koala looks like a teddy bear come to life. It measures about 2½ feet (76 cm) long. Small, brown eyes stare out from its wide, round head. Soft fur covers its chubby body and fringes its ears. Its large nose is smooth and leathery.

Though the koala looks like a bear, it is not a bear at all. It is a pouched mammal called a marsupial (say mar-soo-pea-ul).

Koalas live in eastern Australia. Eucalyptus trees there provide koalas with food and homes. The animals rarely leave the branches of these trees. The koala climbs well. It spreads its toes and grasps the sides of a small branch. Its sharp claws dig into the bark.

Even when the koala moves to a different tree, it does not always come down to the ground. In thick forests, for example, it climbs along a branch toward a nearby tree. When it is close enough to the other tree, the koala jumps.

Koalas spend the day dozing high in the trees. Sometimes they curl up in a tree fork. Or they may sit on a limb with their legs hanging down. Toward evening, koalas become more active. They climb

Clinging fast to its mother's soft fur, a young koala ▷ rides piggyback up a limb. Female koalas carry their offspring in pouches for the first six months.

◁ Perched high in a eucalyptus tree, a koala keeps safe from most enemies except large owls. Koalas seldom come down from such dizzying heights.

along the branches as they feed on eucalyptus leaves and bark. Each koala eats about $2\frac{1}{2}$ pounds (1 kg) of leaves a day, and drinks very little water. The leaves supply most of the moisture a koala needs.

Though they eat a lot for their size, koalas are careful feeders. They may pass up several leafy branches before stopping to eat. At certain times of the year, eucalyptus leaves contain a poisonous acid. Koalas avoid these deadly leaves.

Because koalas need so much food, it is hard to keep them in zoos. A hundred tall eucalyptus trees are needed to provide food for each animal. With fewer trees, koalas soon would strip every branch bare. Neither the trees nor the koalas would survive. Outside Australia, the only place in the world where koalas live is at the San Diego Zoo in California. Many eucalyptus trees have been planted there.

In the wild, koalas usually live alone. During the mating season, however, they often form small groups. One male and several females may stay together. Koalas are generally quiet animals. But during mating season males are very noisy. Their calls sound like saws cutting wood in a forest.

A female koala gives birth to one offspring, usually in the spring or summer. The newborn koala is hairless and no bigger than a grape. It climbs blindly into the protective pouch on its mother's belly. There the tiny koala attaches itself to a nipple and nurses for several months.

Bandicoots, kangaroos, and wombats also bear very small young that develop in pouches. Read about these marsupials under their own headings.

At six months of age, a young koala is strong enough to leave the pouch. But even then it travels everywhere with its mother. The small animal rides on her back, clinging to her fur with its claws. When resting, it hugs her belly. If the young becomes separated from its mother, it cries out for her.

After a young koala has left the pouch, it still returns there for food. For a few more months, it drinks only its mother's milk. Then it begins to eat leaves. By the time a koala is a year old, it no longer needs its mother. Soon it will go off and live alone.

Koala

*Wide yawn and sleepy eyes signal nap time. A koala ▷
spends as much as twenty hours a day dozing and
resting in the trees. It wakes from time to time to eat a
leafy snack. But the animal feeds mainly after dark.*

△ *Snug in a double-deck bunk, two koalas curl up for a
nap. Even swaying treetops will not disturb a sleeping
koala nestled between two branches.*

When it is three or four years old, it will begin to
have offspring of its own.

Koalas once were common in many parts of
Australia. But over the years hunters killed large
numbers of the animals for food and for their thick,
warm fur. To clear the land for farms, people cut
down the trees where the koalas lived. Fires destroyed some of the forests.

As koalas began to die out, Australians became
concerned that the species would soon be extinct.
Strict laws were passed to keep the animals from
harm. Today koalas are increasing in number.

The koala is a member of the phalanger family.
Read about other phalangers on page 436.

Kob
The kob is a kind of antelope. Learn about antelopes on page 52.

Kudu
The kudu is a kind of antelope. Read about antelopes on page 52.

L

Langur
The langur is a kind of monkey. Read about monkeys on page 376.

Lemming

(*say* LEM-ing)

▽ *Tall grass shelters a southern bog lemming in an Ohio field. Lemmings make runways, or paths, by tunneling through grass or under snow.*

Southern bog lemming: 4 in (10 cm) long; tail, less than 1 in (3 cm)

PEOPLE HAVE TOLD MANY TALES about the thousands of lemmings that every so often march into the sea and die. Though the storytellers are exaggerating, they are not simply spinning yarns.

According to scientists, some lemmings do migrate, or travel, long distances. Every three or four years, the population of Norwegian lemmings vastly increases. Then the animals leave their homes and scatter in all directions. Many of them soon die from exhaustion or starvation. Some become the prey of larger animals. Others at last make their way to the ocean's edge. Strong swimmers, some might be able to cross a small body of water. In the ocean, however, they drown.

Experts still do not fully understand why the furry animals make such journeys. Lack of food and of space for the huge numbers of lemmings may force some of the rodents to move elsewhere.

Norwegian lemmings are one of about a dozen kinds of lemmings. Most live in the cold northern areas of Europe, Asia, and North America. Some live in the northeastern and midwestern United States. Lemmings measure about 5 inches (13 cm) from their short snouts to the tips of their stubby tails. Most lemmings have gray or brown fur.

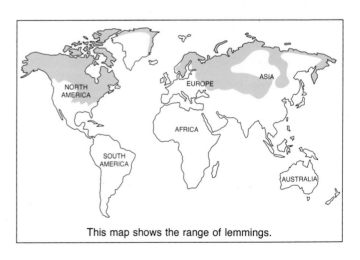

This map shows the range of lemmings.

Norwegian lemming: 5 in (13 cm) long; tail, less than 1 in (3 cm)

▷ *Furry, colorful ball, a Norwegian lemming sits on a bed of green moss. A thick coat hides its small ears and short tail.*

▽ *Carrying her offspring in her mouth, a collared lemming in Canada moves her young from her grassy nest. A female lemming usually has three to seven young in a litter. She may give birth several times a year.*

▽ *Well hidden by its light winter coat, a collared lemming digs in the snow in Alaska. In summer, the rodent's fur turns grayish brown.*

Collared lemming: 5 in (13 cm) long; tail, less than 1 in (3 cm)

Lemmings feed mostly on plants. In summer, these rodents live in burrows under mosses and other plants. They make paths called runways through the grass or under roots. In winter, the animals often live in tunnels under the snow. There they find food as well as shelter from the cold. And they have better protection from such enemies as weasels and foxes.

A female lemming usually gives birth to three to seven young. A pregnancy lasts about three weeks, and females may have several litters a year.

LEMMING

LENGTH OF HEAD AND BODY: 3-6 in (8-15 cm); tail, less than 1 in (3 cm)

WEIGHT: $^1/_2$ oz-4 oz (14-113 g)

HABITAT AND RANGE: meadows, woods, marshes, and tundra in parts of North America, Europe, and Asia

FOOD: mostly plants such as mosses and grasses

LIFE SPAN: usually less than 2 years in the wild

REPRODUCTION: usually 3 to 7 young after a pregnancy of about 3 weeks

ORDER: rodents

Lemur

△ *Hooking its legs over and under the branches, a sifaka, a kind of lemur, curls up in a tree in Madagascar. When resting, the animal often coils its tail.*

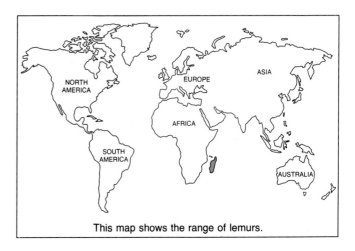

This map shows the range of lemurs.

THROUGHOUT THE FORESTS of Madagascar live large-eyed animals called lemurs. Lemurs are found only on this large island off the southeastern coast of Africa and on the small neighboring Comoro Islands. Lemurs, like apes, monkeys, and humans, belong to the primate order.

Some scientists think that Madagascar was once part of the continent of Africa. Gradually, the

Sifaka (below) soars gracefully through the air. Long hind legs help this lemur spring from one branch to another. At right, sifakas rest in the branches of a tree. On the ground, they sometimes run in an upright position.

Sifaka: 18 in (46 cm) long; tail, 22 in (56 cm)

Black lemur: 10 in (25 cm) long; tail, 15 in (38 cm)

Mouse lemur: 5 in (13 cm) long; tail, 6 in (15 cm)

△ *Eyes shining, black lemurs peer from a perch in the trees. Despite their name, only males have black fur. Females have coats of rusty brown.*

◁ *Among the smallest of all primates, a mouse lemur clings to a narrow limb. The animal looks for food at night. It sleeps all day curled in a hollow tree or in a leafy nest.*

leaping and climbing. They use their hands and feet to grip the branches.

There are many species, or kinds, of lemurs, and they vary greatly in size. The mouse lemur, one of the smallest of the primates, measures only 5 inches (13 cm) long, not including its tail. The indri (say IN-dree) is the largest lemur. Its body measures more than 2 feet (61 cm) long.

Mouse and dwarf lemurs usually live alone and scurry along tree branches at night. Their big ears help them search for food and hear such enemies as catlike fossas. All lemurs eat plants, but mouse and dwarf lemurs also feed on insects.

Both the fat-tailed dwarf lemur and the mouse lemur have tails that are sometimes padded with fat.

island separated from the continent. Because of their isolation, lemurs have little competition for their food and only a few enemies. They are well suited to their environment. Today lemurs remain relatively unchanged from their ancestors.

Many lemurs have slender bodies and narrow, pointed snouts. Lemurs move about the trees by

Brown lemur: 15 in (38 cm) long; tail, 20 in (51 cm)

△ *Brown lemur travels on all fours along the stiff, broad leaf of a coconut palm. The animal's long tail helps it balance. Brown lemurs feed on insects and fruit during the day and at night.*

Mongoose lemur: 14 in (36 cm) long; tail, 19 in (48 cm)

△ *Downy light fur covers the chest of a mongoose lemur. Strong, flexible hands help it grip branches.*

Tail held high, a ruffed lemur strides across a field. ▷ *Usually ruffed lemurs stay in the trees. The white fur under its chin gives the animal its name.*

Ruffed lemur: 22 in (56 cm) long; tail, 24 in (61 cm)

Ring-tailed lemur: 16 in (41 cm) long; tail, 23 in (58 cm)

Troop of ring-tailed lemurs feasts on leaves. These lemurs spend much of the time on the ground.

Lemur

When food becomes scarce, these animals begin to absorb the fat stored in their tails.

Typical lemurs are larger than dwarf and mouse lemurs. They usually grow as big as house cats. Nearly all typical lemurs live in groups. Some kinds, such as the mongoose lemur and the ruffed lemur, stay together in small families. Others, such as the ring-tailed lemur and the black lemur, live in larger groups called troops.

When moving through the trees, typical lemurs walk along the branches on all fours. Often they make long jumps from branch to branch. The ring-tailed lemur, however, spends more time traveling on the ground than do other lemurs.

The indri, the avahi (say uh-VAH-hee), and the sifaka (say suh-FAHK-uh) make up the indri family. Unlike other lemurs, members of this family occasionally stand upright when walking or running. At rest, they sit up and cling to branches. Sifakas often make spectacular leaps. Their long back legs give them a powerful takeoff. With arms and legs outstretched, they may leap 20 feet (6 m) or more.

During the day, the black, hairless snouts and light-colored fur of many sifakas are easy to see. The woolly avahi, however, curls up in trees and sleeps in the daytime. Little is known about the animal in the wild because it is rarely seen by people.

Lemur pregnancies range from two to nearly five months, depending on the kind of lemur. A female may give birth to one, two, or three young. Ruffed lemurs, mouse lemurs, and dwarf lemurs build nests for their offspring. A female ruffed lemur often parks her offspring on a branch while she looks for food. Other lemurs take their young with them. The newborn cling tightly to their mother's fur as she moves among the branches. Young lemurs begin to play as soon as they can walk. At about two years of age, they are independent.

LEMUR

LENGTH OF HEAD AND BODY: 5-28 in (13-71 cm); tail, 5-26 in (13-66 cm)

WEIGHT: 2 oz-13 lb (57 g-6 kg)

HABITAT AND RANGE: forests of Madagascar and the Comoro Islands

FOOD: fruit, flowers, leaves, bark, insects, and tree gum

LIFE SPAN: 20 years in captivity

REPRODUCTION: usually 1 to 3 young after a pregnancy of about 2 to 4½ months, depending on species

ORDER: primates

Leopard

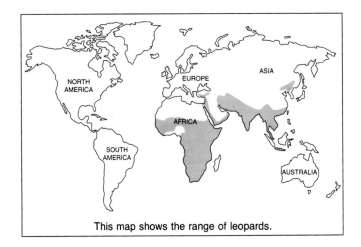

This map shows the range of leopards.

STARING INTENTLY, the leopard pauses quietly in tall grass. It flicks its tail. Then the big cat rushes toward its prey—an antelope. Seizing its victim with its claws, the leopard pulls the animal to the ground. It kills the antelope by biting into the neck or the throat. Instead of eating its meal on the spot, the leopard usually drags it away. It finds a place safe from hyenas, vultures, and lions.

A leopard weighs only about 100 pounds (45 kg). But the muscular cat is so strong that it can carry an animal its own weight up a tree. There it wedges its prey among the branches. The leopard feeds in the tree. Any meat that is left will be used for later meals. Leopards also hunt small animals such as rodents, birds, monkeys, and even fish. They will eat almost anything they can catch.

The leopard hunts mostly at night. Except at mating time, it avoids other leopards. If it sees another leopard, it will usually turn away. Like all cats, it uses many signals to alert other leopards to its presence. Rubbing its cheek against a tree, a leopard leaves a scent on the bark. The leopard may claw the tree trunk, or it may spray the tree with urine.

Leopards also let other leopards know where they are by making sounds. All big cats—jaguars, leopards, lions, and tigers—roar. But a leopard's roar does not sound (Continued on page 332)

Slinking through the grass, a leopard stalks prey in Sri Lanka. Leopards are smaller than lions and tigers and slimmer than jaguars. They live in more climates and habitats than these other big cats do.

Leopard

Body molded to the branches, a leopard lounges in a tree. Its paws and tail dangle. Leopards in

Africa often hide food high among the branches where hyenas, lions, and vultures cannot reach it.

◁ Female leopard carries her month-old cub in her mouth. She gently grasps the loose skin at its neck in her teeth. Leopards sometimes move their cubs from one spot to another. They try to keep them hidden in a safe place such as the base of a hollow tree.

at all like a lion's. The leopard has a call that sounds like a deep, rasping cough.

Leopards live in many parts of Africa and Asia. They can survive in forests and on grasslands, in warm and cold climates. No matter where leopards live, their colors and markings help hide them. Most leopards have yellowish coats with dark spots called rosettes. A few leopards appear to be solid black because they have black rosettes on black backgrounds. People call these leopards black panthers.

Leopard cubs are born with dull gray fur. Their spots are barely visible. Females usually give birth to two cubs. At first, a female leopard hides her cubs in a quiet spot—in a cave or a hollow tree. Later, she

may move the cubs from place to place. When the cubs are older, she will lead them to a kill. The young play by stalking and pouncing. The games help prepare them to hunt on their own. After about two years, when they are almost fully grown, the cubs go off by themselves.

Read about snow leopards and other cats beginning on page 126. Find out about jaguars, lions, and tigers in their own entries.

LEOPARD

LENGTH OF HEAD AND BODY: 41-67 in (104-170 cm); tail, 26-38 in (66-97 cm)

WEIGHT: 66-176 lb (30-80 kg)

HABITAT AND RANGE: forests, open woodlands, scrublands, plains, and mountains in parts of Africa, the Middle East, and Asia

FOOD: antelopes, deer, rodents, birds, monkeys, and fish

LIFE SPAN: as long as 21 years in captivity

REPRODUCTION: 1 to 4 young after a pregnancy of about 3½ months

ORDER: carnivores

△ *Almost hidden, a leopard waits in tall grass in Africa. The animal's spotted coat blends with its surroundings. The leopard creeps very close to such prey as antelopes, deer, and wild pigs. Then it rushes out, pulls its victim to the ground, and kills it with a bite.*

Two young female leopards plunge into a river in India (left). In the shallow water, they tussle and play (above). Strong swimmers, leopards often find food in or near water. They eat fish, crabs, and other water animals.

Linsang

WITH ITS SHARP, CURVED CLAWS, the linsang of Asia and Africa can climb almost anywhere. The animal is a skillful hunter. It looks like a cat, and it can retract its sharp claws. But the linsang is a kind of civet. Read about other civets on page 154.

Dark spots, like the shadows of leaves on a tree limb, mark the linsang's short, light-colored fur. Dark rings circle its long tail. The linsang usually becomes active after dark. It moves silently among the trees. A bird, a rat, or a lizard may not hear the linsang until it is too late.

The African linsang not only hunts in trees, but it lives there as well. Not much is known about the animal. Scientists think that these linsangs build nests of leaves and vines among the branches. The animals sleep there during the day. A linsang may have several nests in its home range.

There are two kinds of Asian linsangs. They live in parts of southern and southeastern Asia. Asian linsangs, like their African relatives, usually live alone. They hunt in forests and near fields. They sleep in hollow trees or among tangled roots.

A female linsang gives birth to two or three offspring each year.

Bold pattern marks the soft fur of a banded linsang of Asia. The spots help hide the animal, a kind of civet, in shadowy forests.

Banded linsang: 16 in (41 cm) long; tail, 14 in (36 cm)

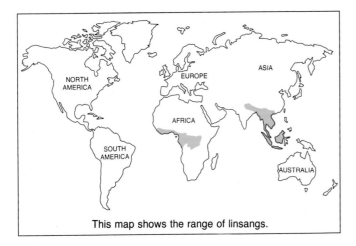
This map shows the range of linsangs.

LINSANG

LENGTH OF HEAD AND BODY: 13-18 in (33-46 cm); tail, 12-15 in (30-38 cm)

WEIGHT: 21-28 oz (595-794 g)

HABITAT AND RANGE: dense forests of western and central Africa and parts of southern and southeastern Asia

FOOD: rodents, insects, reptiles, birds, and fruit

LIFE SPAN: 9 years in captivity

REPRODUCTION: 2 or 3 offspring after a pregnancy of unknown length

ORDER: carnivores

Lion

DUSK SETTLES on the grassy plains of Africa. A pride, or group, of lions begins to stir. The golden cats, each about 6 feet (183 cm) long, rub against one another in greeting. Stretching, they groom themselves. A male shakes his mane, the ruff of hair on his head and neck. Then the lions hear a distant roar—the call of a member of the pride or that of a stranger. The lions respond with more roars.

The roaring is a signal. To pride members it means "Here I am." To other lions it is a warning to stay away. Lions in the same pride keep in touch with a variety of other sounds as well—grunts, meows, growls, and moans.

Lions are the only cats that live in permanent groups. A pride may include as many as six adult males and even more lionesses and cubs—perhaps

Wide yawn reveals the long, sharp teeth of a meat eater. Only an adult male lion has a mane, a ruff of long, thick hair. The bigger the mane, the more impressive a lion looks to other males.

25 animals in all. The lionesses in the pride are all related. Female cubs usually stay with the pride. Male cubs, however, leave when they reach three years of age. Brothers often wander together. Later, they may take over prides from other males.

Other wild cats usually live alone in wooded areas and hunt relatively small prey. Lions, however, live on grassy plains and in open woodlands where herds of large animals roam widely. There, by hunting together, lions can bring down such prey as a wildebeest or a zebra. The kill provides food for many animals. But a hungry lion will hunt whenever it finds prey—even if the lion is alone.

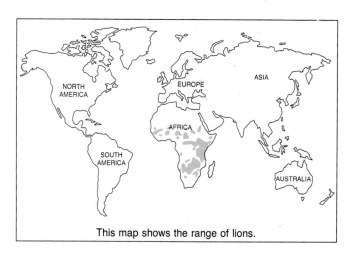

This map shows the range of lions.

The home area of a pride may cover as many as 100 square miles (259 sq km). Within this area, males mark certain places with urine. This lets other lions know that the area is taken. The animals also roar to announce their presence. Within their area, lionesses know the best places to hide cubs and to catch prey. When a strange lion enters another's range, it is chased from the area.

Female lions do most of the hunting for a pride. Because lionesses have no manes, they can stalk their prey unseen more easily than males can.

A pride may rest all day near a herd of antelopes or zebras. At night, the lions begin to stalk the herd. They fan out and circle their prey, drawing as close as possible. When they are within a few feet, one or two of the lionesses may rush at the herd. They may catch a victim themselves, or they may drive it

◁ *Keeping their distance, thirsty zebras wait until a resting lion has left a water hole. Already well fed, the big cat shows no interest in hunting.*

within reach of other lions. Sometimes they may kill more than one of the herd.

Prey is often difficult to catch. Most animals can outrun lions. In some parts of Africa, zebras and wildebeests move away from the plains during the dry months. Lions then must hunt the scattered animals that remain on the plains and in the woodlands. The lions sometimes steal the kills of wild dogs and hyenas. When there is much food, lions gorge themselves. Members of the pride often fight over a kill. Each lion tries to take as much food as possible. When prey is plentiful, cubs get a share. When there is little prey, the cubs may starve.

Newborn lion cubs are covered with spotted, woolly fur. Their eyes usually open after a few days. At first, the cubs are helpless. For the first weeks of their lives, they stay hidden in the grass. Although they usually nurse for about seven months, they may start eating meat at three months of age.

Cubs join the pride when they can move well enough to keep up—after about eight weeks. At that

▽ *Adult lions feed on a wildebeest on a plain in Tanzania. Cubs watch and wait their turn.*

Lion

time, several lionesses bring together their litters—born within a few weeks of each other. A lioness will nurse or guard any cub, not just her own.

The young animals play games with other cubs, learning to be members of a pride. Play strengthens ties between the lions. It also gives cubs practice in hunting. They will not actually take part in a kill for about a year. In the meantime, the cubs wrestle.

They stalk their mothers' tails, pouncing and swatting. One cub may rub against an adult male. Another may sit on its mother's outstretched forelegs.

Lions once roamed most of Africa as well as parts of Europe and Asia. Today scattered groups of lions live in Africa, but only south of the Sahara. In the Gir Forest in India, about 200 lions struggle to survive in an ever shrinking habitat.

▽ *Pride, or group, of lions stretches out lazily under an acacia tree. Lions may rest for twenty hours a day.*

Curled up in tall grass, a lioness in Africa seeks shelter ▷ from the midday sun. The animal's yellowish coat blends well with her habitat.

◁ *Female lion and her six-month-old cubs sit panting in the morning heat after feasting on a zebra.*

▽ *Two-month-old lion cub peers over a rock. The dark spots on a newborn's coat fade gradually as it grows.*

Lion

▽ Four male lions march across a grassland in Africa. A jackal prowls nearby. Lions like these that live peaceably together usually come from the same pride. Four males can hold a larger territory and for a longer period of time than a single male can. Females in the pride are related. They remain in the same area for generations.

△ Hunting alone, a lioness dashes from her hiding place to attack a group

of warthogs. After the kill, members of the pride fight over the meat. Each tries to get as much as it can.

LION

LENGTH OF HEAD AND BODY: 60-72 in (152-183 cm); tail, 24-34 in (61-86 cm)

WEIGHT: 265-420 lb (120-191 kg)

HABITAT AND RANGE: grassy plains and open woodlands in parts of Africa and in the Gir Forest in western India

FOOD: antelopes, zebras, buffaloes, and smaller animals

LIFE SPAN: as long as 15 years in the wild

REPRODUCTION: 1 to 6 young after a pregnancy of 3½ months

ORDER: carnivores

▽ *Swishing her tail, an Asian lioness prowls a forest in search of food. In India, only about 200 of these rare animals survive in the wild. Lions once roamed parts of Europe and Asia and most of Africa.*

Llama

(say LAHM-uh)

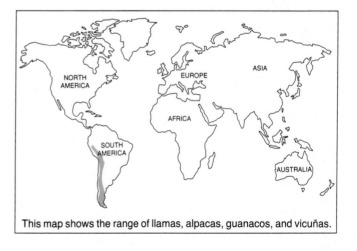

◁ *Ears perked, a woolly young llama—a South American relative of the camel—waits in a dusty field.*

This map shows the range of llamas, alpacas, guanacos, and vicuñas.

Llama: 45 in (114 cm) tall at the shoulder when fully grown

△ Colorful ear tassels dangling, llamas travel a high mountain trail in Peru. The tassels indicate the animals' owners. Llamas provide transportation, food, and fuel for the Indians of South America.

A PACKTRAIN of llamas winds along a rugged mountain trail, carrying goods to a market town. The strong, surefooted animals look like small camels without humps. And they are! Llamas and their close relatives are members of the camel family. They all live in the high Andes of western South America. Llamas and alpacas (say al-PAK-uhs) are now domestic, or tame, animals. Guanacos (say gwuh-NAHK-ohs or wuh-NAHK-ohs) and vicuñas (say vi-KOON-yuhs) live in the wild.

During the day, the animals feed mainly on grasses. They chew their food only a little before they swallow it. After eating, they bring up a wad of the partly digested food, called a cud, and chew it completely. They swallow it again and digest it.

For centuries, Indians of the Andes have used male llamas as pack animals. Female llamas are kept to bear young. The llamas have provided their owners with meat, hides for leather, and wool for ropes, rugs, and other useful objects. Their droppings are burned as fuel.

A llama usually carries a load of 50 to 75 pounds (23-34 kg) strapped on its back. If a pack is too heavy, though, the animal may lie down and refuse to move. Even the shouts and prods of its annoyed owner may not move the stubborn animal. A packtrain may contain several hundred llamas. It can travel as many as 20 miles (32 km) a day.

Alpaca: 39 in (99 cm) tall at the shoulder when fully grown

Blanket of thick, soft wool covers a young alpaca—to the feathery tips of its ears.

LLAMA, ALPACA, GUANACO, AND VICUÑA

HEIGHT: 28-49 in (71-124 cm) at the shoulder

WEIGHT: 100-290 lb (45-132 kg)

HABITAT AND RANGE: mainly dry areas in the Andes and in plains and forests of southern South America

FOOD: grasses, herbs, and other plants

LIFE SPAN: as long as 28 years in captivity

REPRODUCTION: 1 young after a pregnancy of about 11 months

ORDER: artiodactyls

Alpacas look like shaggy, long-haired llamas. Some alpacas' coats grow so long that their woolly hair drags along the ground! South American Indians raise the animals for their soft, thick wool. The Indians shear the wool and sometimes dye it. Then they weave it into a warm, lightweight fabric that sheds rain and snow. This fabric is called alpaca.

Guanacos—the most widespread members of the South American camel family—range from the coast to the mountains. The vicuña is found only in a few areas high in the Andes. Because people hunted these wild animals for their wool, their numbers decreased sharply. Today laws help protect the

△ *Only a few days old but already surefooted, a young alpaca stands next to its mother. If left unsheared, the coats of some alpacas may drag along the ground.*
▽ *Alpacas in Peru stroll past a lake dotted with flamingos. Alpacas graze on high plains during the day. At night, they return to the stone corrals of their owners.*

345

Vicuña: 36 in (91 cm) tall at the shoulder

◁ *Alert male vicuña stands guard over his territory. He shares this area with several females and their young. Vicuñas live only in a few areas high in the Andes.*

▽ *Vicuña family kicks up a cloud of dust as it gallops across a dry lake bed in Peru.*

Guanaco: 42 in (107 cm) tall at the shoulder

△ *Family group of guanacos races across a plain in Argentina. If chased by such enemies as mountain lions, the guanacos can reach speeds of nearly 35 miles (56 km) an hour. The male usually stays in the rear, allowing females and young to escape first.*

animals. Vicuña wool is among the finest in the world. Centuries ago, only Indian royalty could wear clothing made from this wool.

Guanacos and vicuñas sometimes graze near herds of domestic alpacas and llamas. The wild animals live in family groups or in bachelor herds. Young males without territories wander in herds of as many as 150 animals.

Family groups include a male, about six females, and their young. They roam together in a territory, or area, defended by the male. He mates with the females in the group. The animals often feed in the valleys. They sleep on the ridges and slopes. If attacked by such enemies as mountain lions, male vicuñas and guanacos sound a warning. They usually bring up the rear as the group runs away.

About 11 months after mating, female llamas, alpacas, guanacos, and vicuñas each bear a single young. When young vicuñas and guanacos are about one year old, they are chased away from the group by the male. A young male will join a bachelor herd. A female will wander off in search of a new group. Because llamas and alpacas are domestic animals, they remain in their owners' herds.

Loris

Slow loris: 15 in (38 cm) long; tail, about 1 in (3 cm)

Clutching a thin branch, a slow loris rests in a mountain forest in India. Large, round eyes help the loris find food at night. With a quick, powerful motion, the animal easily grabs such prey as lizards and large insects.

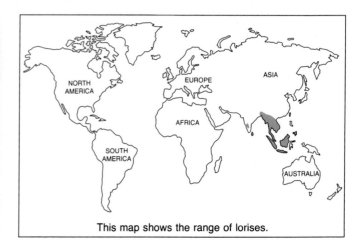

This map shows the range of lorises.

CREEPING AMONG THE TREES after prey, the loris moves along slowly. At each step, it grasps a branch tightly with its broad, muscular feet and hands. The loris inches forward first with one hand and then with the opposite foot. It seems to be traveling in slow motion. It moves so carefully that even the leaves do not rustle as it passes by. The loris can sneak up on a lizard or a grasshopper—then quickly grab its prey. Enemies such as leopards and civets usually do not notice the grayish brown loris climbing in the trees.

Lorises are found deep in the tropical forests of Asia. They rarely come down to the ground. These squirrel-size animals belong to the primate order, which also includes monkeys, apes, and humans. The loris is a relative of the potto, which lives in Africa. Read about the potto on page 459.

Lorises have pointed snouts and small, triangular faces. Their eyes are large and round, and small ears are hidden away in thick, coarse fur. Their tails are short and stubby.

Lorises usually live alone. Occasionally, however, they roam in pairs or in small family groups. The animals hunt after dark and sleep during the day, rolled up in hollow trunks or in forks of trees. A sleeping loris looks like a ball of fur. It tucks its head and arms between its legs and curls up.

The loris can grip tightly with its hands and its feet, even when it sleeps. It can dangle from a tree limb by its feet and use its hands to hold fruit or plants to nibble on. Sometimes lorises hang from branches to stretch, to cool off in hot weather, or to play with each other. When a loris wants to take a drink, it sometimes finds a leaf covered with raindrops or dew. The animal touches the wet leaf and sucks the moisture from its fingers.

Some lorises have long arms and legs and very thin bodies. They are called slender lorises, and they live in the forests of India and Sri Lanka. Slender lorises measure about 10 inches (25 cm) long. The slow loris—a larger, stockier animal—also lives in India and as far east as Indonesia. It may grow as long as 15 inches (38 cm).

After a pregnancy of about six months, a female loris gives birth to a single offspring or sometimes to twins. The young are born with their eyes open, and they are covered with a thin layer of fine fur.

A young loris often clings to the fur on its mother's belly as she sleeps during the day. At night, a female loris parks her offspring on a branch while she searches for food. She cares for her young for about a year, until it is grown. Then the young goes off into the forest alone.

LORIS

LENGTH OF HEAD AND BODY: 7-15 in (18-38 cm); tail, about 1 in (3 cm)

WEIGHT: 10 oz-4 lb (284 g-2 kg)

HABITAT AND RANGE: tropical forests in southern and southeastern Asia

FOOD: lizards, insects, fruit, and tender shoots

LIFE SPAN: as long as 13 years in captivity

REPRODUCTION: usually 1 young after a pregnancy of about 6 months

ORDER: primates

Slender loris: 10 in (25 cm) long; tail, about 1 in (3 cm)

△ *Slender loris creeps carefully along a tree branch. The long-limbed animal moves one hand or foot at a time. Slender lorises live in southern India and in Sri Lanka.*

◁ *Young slow loris hangs on tightly with hands and feet and climbs down a tree headfirst. As it descends, the animal makes a high-pitched, chattering sound.*

349

Lynx

DURING WINTER, thick fur protects the lynx in its home throughout much of the Northern Hemisphere. Even the bottoms of a lynx's large paws are covered with fur. The furry toes spread and make it easier for the animal to walk in deep snow.

The lynx's fur varies greatly in color. It can be light brown, gray, or a shade in between. Sometimes the fur is marked with spots of dark brown. Because of the richness and beauty of the fur, people have hunted the lynx for hundreds of years. As a result of hunting and of destruction of the lynx's habitat, the animals now remain only in scattered areas in Europe. They also live in Asia, in Canada, and in Alaska. The lynx is found on more continents than any other kind of wild cat.

Lynxes are short-tailed cats like their close relatives, bobcats. But the lynx has longer legs, larger paws, and bigger tufts of hair on its ears. The lynx also has much longer side-whiskers on its face than the bobcat does. Read about the bobcat on page 104.

Like most wild cats, the lynx lives alone in a home range. Though it thrives in forests, it easily adapts to dry scrubland and rocky hillsides. It lives

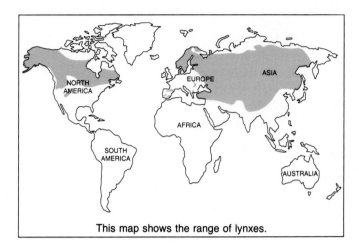

This map shows the range of lynxes.

in areas undisturbed by people—wherever it can find prey. In North America, lynxes feed on mice, squirrels, birds, and especially on snowshoe hares. In Europe, where lynxes are about twice as heavy, they hunt deer as well as smaller prey.

About every ten years, the number of snowshoe hares in North America declines. The number of lynxes then also decreases. Because of lack of food, lynxes give birth to fewer young. The offspring that are born often starve. Normally, a female lynx bears one to six young in the spring.

LYNX

LENGTH OF HEAD AND BODY: 31-51 in (79-130 cm); tail, 4-9 in (10-23 cm)

WEIGHT: 13-66 lb (6-30 kg)

HABITAT AND RANGE: forests, scrublands, and rocky hillsides in Europe, Asia, and northern North America

FOOD: hares, rabbits, rodents, deer, and birds

LIFE SPAN: probably less than 10 years in the wild

REPRODUCTION: 1 to 6 young after a pregnancy of about 2 months

ORDER: carnivores

◁ *Tracks lead to a Canada lynx in a clearing. In winter, the broad, furry paws of the animal serve as snowshoes and help keep the lynx from sinking in deep snow.*

Soft coat of spotted light brown fur covers the body ▷ *of a Eurasian lynx. These animals—prized for their coats by hunters—once lived throughout much of Europe.*

Canada lynx: 34 in (86 cm) long; tail, 4 in (10 cm)

Eurasian lynx: 46 in (117 cm) long; tail, 7 in (18 cm)

M

Macaque
The macaque is a kind of monkey. Read about monkeys on page 376.

Manatee
(*say* MAN-uh-tee)

In the clear waters of Florida's Crystal River, a West ▷
Indian manatee moves away from two divers. In places
where divers often swim, some manatees seek attention
from people. Most avoid people, however. In past years,
manatees were hunted for food, for hides, for oil, and for
bones. Today laws protect them.

West Indian manatee: 13 ft (4 m) long

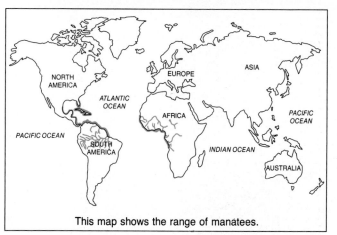

This map shows the range of manatees.

IT'S HARD TO BELIEVE that sailors once mistook
manatees for mermaids. These sea mammals look
more like small blimps than beautiful women. Manatees grow much larger than people. Some reach 13
feet (4 m) in length and weigh as much as 2,000
pounds (907 kg). Manatees are harmless. They rarely fight each other, and they have no natural
enemies. Manatees are as calm as cows grazing in
a meadow. Sometimes they are called sea cows.

There are three kinds of manatees, but all of
them look and behave much the same. They all stay
in warm waters. One kind ranges from Florida to the
coastal areas of Brazil. Another inhabits the Amazon

◁ *Graceful in its watery world, a West Indian manatee*
rises to the surface to breathe. Although it spends all its
life in the water, a manatee comes up for air every few
minutes while swimming. A resting manatee can stay
underwater for as long as 15 minutes.

353

River. The third kind is found along the western coast of Africa.

Though the manatee is about as long and as heavy as a subcompact car, it is a graceful and agile swimmer. Manatees swim alone, in pairs, or in small groups of three to six animals. Usually they move in slow motion. They cruise, or swim at a steady pace, at 5 miles (8 km) an hour. But they can cover short distances at 15 miles (24 km) an hour.

When a manatee cruises, it keeps its flippers by its sides. Its strong tail strokes up and down, pushing the animal forward. A manatee's flippers are flexible and useful. They help it steer. In shallow water, manatees use their flippers to walk on the bottom. As they move, they slowly place one flipper in front of the other.

Boaters on the surface of the water usually see only the nose of a swimming manatee. When the

Manatees surface to breathe at a warm spring in Florida. In the fall, when their river home becomes chilly, about 25 manatees migrate to these warmer waters. They will stay there until March.

animal rises to take a breath, most of its body stays underwater. Cruising manatees breathe noisily every three or four minutes. When they are not active, they can stay submerged for as long as 15 minutes.

Manatee calves are born underwater. Each calf must come up immediately to breathe air. The mother helps her newborn reach the surface for the first time. Usually the calf can swim alone within one hour. A manatee weighs from 25 to 60 pounds (11-27 kg) at birth.

After a few months, a calf begins to graze, though it still drinks milk from its mother. The two stay close together for as long as two years—

MANATEE

LENGTH OF HEAD AND BODY: 8-13 ft (244 cm-4 m)

WEIGHT: 440-2,000 lb (200-907 kg)

HABITAT AND RANGE: rivers, bays, and coastal areas from the southeastern United States to central South America, including the Caribbean Sea, and western tropical Africa

FOOD: water plants, sea grasses, and algae

LIFE SPAN: 40 years in the wild

REPRODUCTION: usually 1 young after a pregnancy of as long as 13 months

ORDER: sirenians

▽ *Lying on its back, flippers folded on its chest, eyes closed, a manatee naps upside down. These animals spend about one-third of the day resting.*

cruising, resting, and coming to the surface to breathe. Calves often play with other manatees. They squeak, squeal, and scream. They nuzzle, nibble, and nudge one another. They even embrace with their flippers. Manatees probably recognize each other by the sounds they make and by touch. Their skin is thick and wrinkled but very sensitive.

Manatees feed on plants that grow in the water

▽ *Adult manatee feeds on stringy water plants. Manatees eat huge amounts of food—about one-tenth of their weight every 24 hours. The animals help clear clogged waterways by eating plants that grow there.*

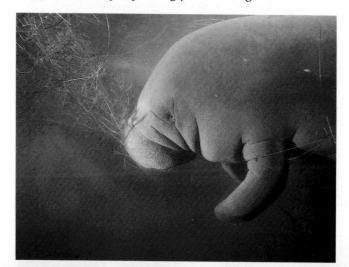

Breathing with its nostrils just above the surface, a manatee hangs ▷ *suspended in the water. Bristles on its upper lip help the animal eat. With these whiskers, it pushes food toward its mouth.*

and at the water's edge: water weeds, sea grasses, and algae. Every 24 hours, a manatee consumes as much as 1 pound (½ kg) of food for every 10 pounds (5 kg) of its body weight. A human child weighing 80 pounds (36 kg) would have to eat 8 pounds (4 kg) of salad a day to keep up with a manatee!

People have hunted manatees for food and for their hides, oil, and bones. In the past century, the number of manatees has slowly decreased. As a result, laws have been passed to help protect the animals. But manatees are often harmed by motorboats that travel through the waters where they live. Many of the slow-moving animals have scars on their bodies caused by propeller blades.

The manatee is related to another sea mammal, the dugong. The animals look and behave much alike. But they inhabit completely different parts of the world. Read about the dugong on page 186.

Mara

(*say* muh-RAH or MAH-ruh)

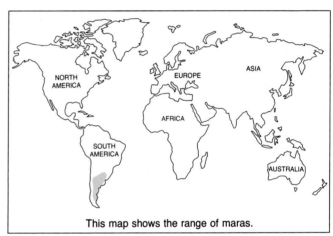

This map shows the range of maras.

Sitting up, a female mara nurses her young.
In this position, she can watch for enemies such as foxes.
A mara's long ears help pick up distant sounds.

BOUNCING ACROSS THE DRY PLAINS of South America, the long-legged mara looks almost like an antelope—but a very small one! It measures only about 30 inches (76 cm) from nose to stubby tail. The mara is actually a large rodent—a relative of the guinea pig. It lives in Patagonia, a dry, windy area in the southern part of South America.

Because of its long ears, some people call the mara the Patagonian hare. Maras are not related to hares, however.

When a mara rests, it sometimes lies with its long front legs tucked under its chest. It grooms itself carefully, wiping its face with a foreleg. As it sits in the sun, the mara flicks its ears. It stays alert to every sound, ready to escape if an enemy such as a fox threatens. Maras can run quickly. As they dash away, patches of white hair on their rumps may serve as danger signals to other maras.

Maras can run across the pebbly ground of Patagonia without being hurt. The bottoms of their feet are protected by hair and thick pads.

The rodents are well adapted, or suited, to their rugged homeland in other ways, too. Maras rarely drink water. They get most of the moisture they need from plants. Their thick coats of brown-and-gray hair blend with the colors of their habitat.

During the day, maras feed on grass and on the scrubby plants of the Patagonian plains. By gnawing and chewing, the animals keep their teeth from growing too long. Like the front teeth of all rodents, a mara's front teeth never stop growing.

Maras dig burrows in which to raise young. A female bears one to three offspring. The young have hair and teeth at birth, and their eyes are open. A female mara nurses her young sitting up. In this position, she can watch for danger.

Young maras follow their mother across the dry plains ▷ *of Patagonia in South America. Born with hair and open eyes, offspring can walk soon after birth.*

▽ *Five small maras run into their burrow for shelter. A sixth has already dashed underground. Several adult females may leave their offspring in one burrow and go off to feed. They come back to nurse the young.*

MARA

LENGTH OF HEAD AND BODY: 24-29 in (61-74 cm); tail, as long as 2 in (5 cm)

WEIGHT: 20-35 lb (9-16 kg)

HABITAT AND RANGE: dry grasslands and scrubby plains in parts of southern South America

FOOD: grass and other plants

LIFE SPAN: about 10 years in captivity

REPRODUCTION: 1 to 3 young after a pregnancy of about 3 months

ORDER: rodents

Marmoset

The marmoset is a kind of monkey. Read about monkeys on page 376.

Marmot

(*say* MAR-mut)

BY THE END OF SUMMER, marmots usually are very fat animals. For several months, they have stuffed themselves with food. They have eaten grasses, leaves, flowers, fruit, and—once in a while—a grasshopper or a bird's egg.

By early fall, the large rodents—about 2 feet (61 cm) long including their tails—have begun to waddle when they walk. Their furry bellies often drag the ground. When a marmot sits up on its hind legs, it looks like a chubby little man.

When the weather turns cold, marmots retreat into burrows. They sleep for six months or more. The body temperatures drop, and the heart rates slow down. This kind of sleep is called hibernation (say hye-bur-NAY-shun). While they hibernate, marmots live on the fat in their bodies.

Marmots are found in parts of Europe, Asia, and North America. They make their homes near fields, in woods, in high valleys, on grassy mountain slopes, and among rocks.

Most species, or kinds, of marmots live in groups called colonies. They eat and sleep in a large underground burrow that they share.

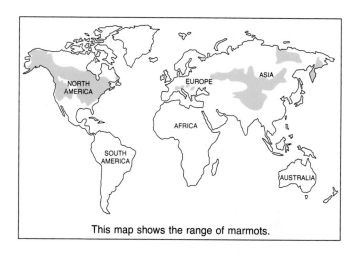

This map shows the range of marmots.

The woodchuck, a North American marmot, behaves differently. This animal—often called a groundhog—lives alone most of the year. It stays with other woodchucks for only a few months, when there are young. The woodchuck usually digs its burrow in a wooded area or in a field. A burrow may extend more than 40 feet (12 m).

Other marmots, such as the hoary marmot and

▽ *Its mouth full of grass, a hoary marmot in Alaska heads for its burrow. Marmots make grassy nests inside their burrows. There they hibernate, or sleep deeply, during the winter—for six months or more.*

△ *Young hoary marmot greets an adult by nuzzling it. Marmots often live in large family groups called colonies. They play and sun together.*

◁ *Hoary marmot picks its way along rocks in Montana. The rodent always remains alert to danger.*

Hoary marmot: 19 in (48 cm) long; tail, 8 in (20 cm)

Marmot

the yellow-bellied marmot, often dig their burrows under boulders on mountainsides. There they are safe from most animals. The yellow-bellied marmot is often called the rockchuck because it lives in rocky parts of western North America.

Marmots mate in the spring. Females usually bear two to six young a few weeks later. The young begin to wander outside the burrow after about a month. But they usually stay with their parents and hibernate with them during their first winter.

Young marmots play together much of the time. They roll and tumble while one family member

keeps watch. Their play helps the marmots learn to live in the colony. Marmots are friendly with each other. A young animal often greets an adult by nuzzling it and touching it with its paws.

Marmots have good eyesight, and they hear well. At the sight or sound of a bear, a fox, or an eagle, a marmot usually gives a high-pitched alarm call. All the marmots then run for cover.

Marmots are the largest members of the squirrel family. Read about their relatives—chipmunks, prairie dogs, and squirrels—in their own entries.

MARMOT

LENGTH OF HEAD AND BODY: **12-24 in (30-61 cm); tail, 4-10 in (10-25 cm)**

WEIGHT: **7-17 lb (3-8 kg)**

HABITAT AND RANGE: **fields, valleys, mountains, plains, and wooded areas in parts of North America, Europe, and Asia**

FOOD: **grasses, leaves, flowers, fruit, and sometimes insects and birds' eggs**

LIFE SPAN: **as long as 20 years in captivity**

REPRODUCTION: **2 to 6 young after a pregnancy of about 1 month**

ORDER: **rodents**

△ *Hoary marmot licks sap that oozes from a willow branch in early spring. Marmots eat mainly plants, including leaves and grasses.*

◁ *Hoary marmot (above, left) basks in the sun on a rock in Alaska. Hoary marmots often dig their burrows on rocky slopes of mountains. There they can lie on boulders and keep watch for danger. Below left, another hoary marmot peers cautiously from its burrow.*

Woodchuck in North Carolina ▷ *sits up to eat a tomato. Like other marmots, woodchucks often hold their food in their forepaws. These animals, also called groundhogs, give their name to Groundhog Day, on February 2. According to tradition, if a groundhog sees its shadow on that day, there will be six more weeks of winter.*

Woodchuck: 17 in (43 cm) long; tail, 5 in (13 cm)

September sunshine bathes an Alpine marmot sitting on a mountain ▷ *slope in Switzerland. Soon it will go underground and hibernate. Its breathing will slow down, and its heart rate will drop. For energy, the plump animal will use the fat stored in its body.*

Alpine marmot: 22 in (56 cm) long; tail, 6 in (15 cm)

Marsupial mouse

(*say* mar-soo-pea-ul mouse)

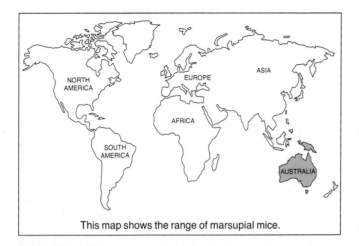

This map shows the range of marsupial mice.

Narrow-footed marsupial mouse: 4 in (10 cm) long; tail, 3 in (8 cm)

△ *On the lookout, a narrow-footed marsupial mouse watches for prey in a forest in Australia.*
◁ *Jerboa marsupial mouse balances on a rock. Like the jerboa, a rodent found in parts of Asia and Africa, the marsupial mouse can bound rapidly across the ground. Though the animals look alike, they are not related.*

Jerboa marsupial mouse: 4 in (10 cm) long; tail, 5 in (13 cm)

WITH THEIR BIG EARS, pointed noses, and long tails, marsupial mice certainly look like field mice. The animals are not related, however. Field mice are rodents. Marsupial mice are members of the same group as kangaroos, koalas, and opossums.

Marsupials give birth to tiny, underdeveloped young. The offspring usually crawl into a pouch on their mother's belly soon after birth. They stay there for several weeks, nursing and growing larger and stronger. The narrow-footed marsupial mouse, for example, has a well-developed front-opening pouch. The jerboa marsupial mouse has a rear-opening pouch. Other marsupial mice may have no pouch at all—or they may have just a small flap of skin around the nipples. Some marsupial mice bear as many as 12 young at one time.

Unlike field mice, which eat mostly plants, marsupial mice eat small animals, including insects. There are about thirty kinds of marsupial mice. They live in all kinds of habitats in Australia, New Guinea, and on nearby islands.

MARSUPIAL MOUSE

LENGTH OF HEAD AND BODY: 2-9 in (5-23 cm); tail, 2-9 in (5-23 cm)

WEIGHT: $1/_5$ oz-6 oz (6-170 g)

HABITAT AND RANGE: all kinds of habitats throughout Australia, New Guinea, and neighboring islands

FOOD: insects, rodents, and other small animals

LIFE SPAN: about 4 years in captivity

REPRODUCTION: 3 to 12 young after a pregnancy of about 3 to 7 weeks, depending on species

ORDER: marsupials

Marten

American marten: 16 in (41 cm) long; tail, 8 in (20 cm)

American marten peers from a branch in the Adirondack Mountains of New York.

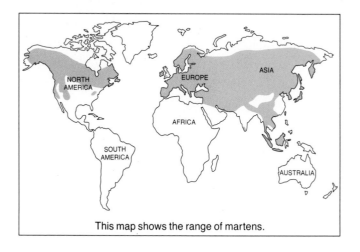

This map shows the range of martens.

LIKE A TIGHTROPE WALKER in an evergreen forest, an American marten scurries along a high branch after a red squirrel. Martens, members of the weasel family, seem to be always on the move. The animals bound across the snow in winter. They dash through hollow logs or scamper through brush. Martens hunt alone. They sniff the air for scents and listen for the rustling noises of small animals. When they spot their prey, they pounce.

Two kinds of martens—fishers and American martens—live in Canada and the northern United States. Beech martens and pine martens live in Europe and Asia. The sable, another kind of marten, lives in Asia.

Martens are about 2 feet (61 cm) long from nose to tip of tail. They grow thick coats in winter. For centuries, people have hunted the animals for their fur. Some of the most expensive fur coats are made of the skins of martens, especially sables and fishers.

In North America, the marten is at home on the ground or in the trees. It may mark its territory, or area, by leaving traces of a strong-smelling substance produced by glands in its body.

American martens and fishers hunt hares, birds, and rodents. They may kill more than they can eat. They hide the food and return to eat it later. Martens may also eat insects and fruit.

From late morning to late afternoon, American martens, sometimes called pine martens, rest. They

363

Fisher: 22 in (56 cm) long; tail, 14 in (36 cm)

△ *In the woods of northern Minnesota, a fisher, a kind of marten, tests a branch. Larger and faster than American martens, fishers usually hunt larger prey. They can even kill porcupines, despite the sharp quills.*

◁ *Light snow clings to the fur of an American marten in Wyoming. Skillful climbers, martens live in parts of North America, Europe, and Asia.*

Young beech martens, sometimes called stone martens, ▷ play near a tree in Germany.

become active in the evening and at night. The fisher hunts by day as well as by night. It is larger than the American marten and usually goes after larger prey. The fisher is one of the few animals that can kill and eat a porcupine. It attacks a porcupine from the front, biting the animal's face again and again. Finally, the fisher flips the porcupine onto its back. It bites into its underbelly, where there are no quills.

American martens and fishers often bear their young in holes in trees. The American marten may sometimes use a hollow log. A female gives birth to

Beech marten: 18 in (46 cm) long; tail, 9 in (23 cm) when fully grown

365

her young—usually three kits—in the spring. By about four months of age, the kits have begun to learn to hunt. In the fall, they are able to leave home.

Beech martens, sometimes called stone martens, rarely go into the trees, though they are good climbers. Most of the time, these animals scamper along the rocky ground looking for rodents, young hares, birds, and berries. Beech martens may live near people—in parks and gardens, in cities and villages. They sometimes catch rabbits and chickens in farmyards. Usually brownish in color, beech martens have lighter underfur that shows through their longer, darker outer fur.

The sable has a coat of thick, silky fur that ranges from the color of straw to nearly black. Sables live in forests and feed on small rodents, nuts, and berries. Though they stay mostly on the ground,

sables may climb trees to escape enemies or to look for food. Sables mate in summer and give birth the following spring. The kits are blind and covered with light-colored fur. After a month, their eyes open. By July, they are ready to go off on their own.

MARTEN

LENGTH OF HEAD AND BODY: **15-28 in (38-71 cm); tail, 5-15 in (13-38 cm)**

WEIGHT: **less than 2-12 lb (1-5 kg)**

HABITAT AND RANGE: **forests and rocky areas in North America, Europe, and Asia**

FOOD: **small mammals, birds, fruit, berries, insects, eggs, and the remains of dead animals**

LIFE SPAN: **9 to 17 years in captivity, depending on species**

REPRODUCTION: **1 to 6 young after a pregnancy of 8 to 12 months, depending on species**

ORDER: **carnivores**

Red persimmons make a tasty meal for a Japanese marten on a nighttime search for food.

Japanese marten: 25 in (64 cm) long; tail, 8 in (20 cm)

Meerkat

Meerkat is another name for the suricate. Read about the suricate on page 532.

Mink

American mink: 18 in (46 cm) long; tail, 6 in (15 cm)

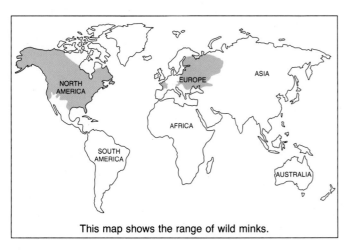

This map shows the range of wild minks.

MINK

LENGTH OF HEAD AND BODY: 13-22 in (33-56 cm); tail, 6-7 in (15-18 cm)

WEIGHT: as much as 3 lb (1 kg)

HABITAT AND RANGE: streams, lakes, rivers, marshes, and swamps in parts of North America, Europe, and western Asia

FOOD: small mammals, frogs, shellfish, fish, eggs, insects, and water birds

LIFE SPAN: 8 to 10 years in captivity

REPRODUCTION: 2 to 6 young after a pregnancy of about 2 months

ORDER: carnivores

On an ice-covered stream, an American mink pauses while looking for food. Besides hunting, minks find places to nest among roots and fallen logs at the water's edge.

SLENDER, SHORT-LEGGED members of the weasel family, minks make their homes along the edges of lakes and streams. There they find food and shelter in thickets, in rock crevices, and among tree roots. These animals—known for their thick, shiny fur—live from Florida into the Arctic in North America. Some are found in Europe and western Asia.

With their partly webbed feet, minks are good swimmers and divers. They slip into and out of the water, looking for fish, crayfish, and frogs to eat. Their prey includes muskrats, hares, and mice. Minks also eat insects and water birds. They may kill more than they can eat at one time. They store what is left and return to it later.

A mink hunts alone, and it seems to be constantly on the prowl. The animal marks its territory, or area, with a strong-smelling substance produced in glands under its tail. These scent marks warn other minks to hunt somewhere else. Minks defend themselves fiercely. When a large owl or a bobcat threatens, a mink fluffs up its fur, hisses, and rushes at its enemy.

Minks mate in late winter or early spring. About two months later, the female makes a nest in a hollow log. Or she may burrow into a riverbank. She bears two to six kits. The young nurse until they are about five weeks old. They play, tumble, and chase each other. After a few months, they begin to hunt with their mother. In the fall, they go off and seek their own territories.

For hundreds of years, people have trapped minks and used their skins for clothing. Today the animals are raised on farms called ranches. Wild minks usually have coats in a shade of brown. But a mink bred on a ranch may have one of many colors of fur—from pure white to jet black.

Mole

IN THE DARKNESS of its burrow, the mole eats, sleeps, mates, and raises young. This nearly sightless, chipmunk-size animal spends most of its life underground. About twenty kinds of moles live in woodlands and fields and along riverbanks in parts of Europe, Asia, and North America. Distant relatives of moles called golden moles live in Africa. There are about twenty kinds of golden moles.

To make its burrow, a mole tunnels through the

European mole: 5 in (13 cm) long; tail, 1 in (3 cm)

This map shows the range of moles and golden moles.

Dirt clinging to its head and body, a European mole (below) surfaces from underground. The shovel-like front feet of this common mole seem to stick straight out from its shoulders. Even when a mole leaves its dark burrow (above), it still cannot see. It has tiny, weak eyes hidden by fur.

ground. For its nest, it digs out a small room. There the mole sleeps on a bed of grass. In passages that lead from the nest, the mole hunts for food.

A mole's body is well adapted, or suited, for digging. Its powerful front feet are big. They seem to grow right out of the animal's shoulders. At the end of each foot are thick claws. As a mole digs, it looks as if it is swimming. First one foot and then the other moves into the earth, pushing back the soil. The mole twists its body forward and presses loose dirt into the walls of the tunnel. Occasionally, it digs up through the surface of the ground. It pushes extra dirt out of this opening, forming a molehill on the surface.

A mole can scurry through its dark burrow backward as well as forward. Its body is tapered at both ends. Its ears are simply fur-covered holes in its head. Thick, velvety fur on its body lies flat, whichever way it is brushed by the walls of a tunnel.

Most kinds of moles depend mainly on their sense of touch. The whiskers on a mole's face and the hairs on its tail and feet help it find its way in the dark. The tip of its long, narrow snout is covered with many tiny bumps. Each one is extremely sensitive. One kind of mole — the star-nosed mole — has a snout with 22 finger-like feelers. These feelers wave and stretch in all directions. They help the mole find food.

△ Still hairless three weeks after birth, young European moles fill a grassy nest. A female European mole gives birth to three or four young each year.

◁ Webbed feet and a long, powerful tail make a desman an excellent swimmer. This Russian mole uses its long, flexible snout to hunt for food underwater. Largest of all moles, the desman pokes its snout above water when it needs to breathe.

MOLE

LENGTH OF HEAD AND BODY: 2-9 in (5-23 cm); tail, 1-9 in (3-23 cm)

WEIGHT: $1/_3$ oz-6 oz (9-170 g)

HABITAT AND RANGE: fields, woodlands, riverbanks, and deserts in parts of Europe, Asia, North America, and Africa

FOOD: insects, earthworms, mice, fish, and other small animals

LIFE SPAN: about 3 years in the wild

REPRODUCTION: 1 to 7 young after a pregnancy of about 1 month, depending on species

ORDER: insectivores

Mole

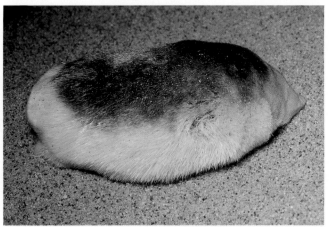

Desert golden mole: 6 in (15 cm) long

Star-nosed mole: 4 in (10 cm) long; tail, 3 in (8 cm)

△ *Desert golden mole, distant relative of the mole, scurries across sand in southern Africa.*

△ *Waving the feelers on its snout, a star-nosed mole emerges from a snowbank in northern Minnesota.*

A mole's body burns energy quickly, and the animal eats its weight in food each day. Every three or four hours, a mole searches for food. Besides worms and insect larvae, it gobbles spiders, lizards, and mice. Some moles dig out passages that end underwater in ponds or streams. There moles can catch insects, fish, shellfish, and frogs.

Moles usually live alone. They meet only during the mating season. Four to six weeks after mating, females bear as many as seven offspring.

Mole rat

Like a wrinkled sausage with teeth, a naked mole rat rests in a scientist's hand. This small rodent lives in underground colonies in parts of eastern Africa.

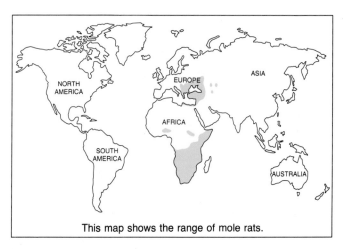

This map shows the range of mole rats.

BURROWING THROUGH THE GROUND, the mole rat chisels with its large front teeth. It packs the walls of its tunnel by pushing with its snout. Its legs swiftly kick dirt toward the entrance.

There are about fifteen kinds of mole rats. They

370

live in parts of Europe, Africa, and Asia. Though these animals live underground like moles, they are really rodents. Their front teeth keep growing, like the front teeth of all rodents. Mole rats gnaw on roots and bulbs and wear down their teeth.

Most mole rats have stocky bodies with gray, black, or brown hair. Their ears are very small, and their tails are short. The lesser mole rat and the greater mole rat seem to have no eyes at all. Actually, their eyes are hidden under the skin. They feel their way with the sensitive hairs on their heads.

The naked mole rat has only a few hairs on its body. Unlike most mole rats, which live alone or in small family groups, naked mole rats live in large colonies. They use teamwork to dig and repair their tunnels, to gather food, and to take care of young. Apparently, only one female naked mole rat in a colony gives birth. Like a queen bee, she bears all the young—as many as 12 in a litter.

MOLE RAT

LENGTH OF HEAD AND BODY: 3-13 in (8-33 cm); tail, as long as 3 in (8 cm)

WEIGHT: 1-53 oz (28-1,503 g)

HABITAT AND RANGE: plains, forests, deserts, and some mountainous areas in parts of Europe, Africa, and Asia

FOOD: mainly roots and bulbs

LIFE SPAN: as long as 8 years in captivity

REPRODUCTION: 1 to 12 young after a pregnancy of 1 to 2½ months

ORDER: rodents

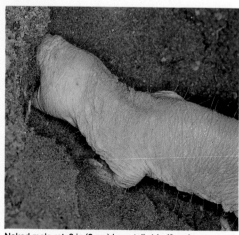

◁ *Flying dirt and a mound of earth give away the presence of a naked mole rat digging a burrow.*
▽ *Naked mole rat sniffs the ground (below, left). Another tunnels through dirt with its large front teeth. Naked mole rats often cooperate when they dig. One mole rat digs and kicks dirt to animals in a line behind it. Each mole rat in turn receives its load of dirt and kicks it backward to the opening. It returns for another load by scampering over the other mole rats.*

Naked mole rat: 3 in (8 cm) long; tail, 1 in (3 cm)

Mongoose

(*say* MONG-goose)

◁ *Stretching its sleek body, a small Indian mongoose watches for eagles and hawks. It uses its pinkish tongue to clean its face after eating.*

▽ *Gray mongoose bites the twisting body of a snake. A group of mongooses sometimes attacks a small snake. Then they fight among themselves for the food.*

Gray mongoose: 16 in (41 cm) long; tail, 13 in (33 cm)

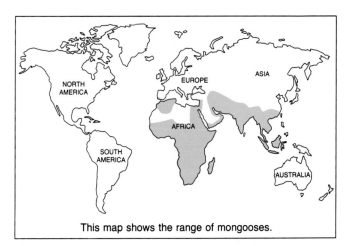

This map shows the range of mongooses.

HAIR BRISTLING on its graceful, slender body, a mongoose circles a cobra. The animal lunges at the poisonous snake. The cobra strikes back. The mongoose dodges and tempts the snake again. With each strike, the snake tires a little more. Finally, the mongoose pounces on the weary cobra. It grabs the snake in its teeth and holds on. Then the mongoose kills the cobra by biting it on the head.

Small Indian mongoose: 12 in (30 cm) long; tail, 9 in (23 cm)

Slender mongoose pauses watchfully on a fallen log in ▷ *Africa. These animals run easily along branches. They usually live alone in scrublands or in open woodlands.*

Such battles usually are staged for entertainment on streets in India. In the wild, however, snakes make up only a small part of a mongoose's diet. Most often, mongooses eat whatever food happens to be nearby. Some eat mostly insects. Others hunt larger prey such as rodents and birds. Many eat fruit and other plants.

Just as mongooses eat many kinds of foods, they live in many places—Asia, Africa, and parts of southern Europe. People have taken mongooses to Hawaii and to the West Indies where they have thrived. Some mongooses make their homes on open, grassy plains or in dense forests. Some are found in swampy areas or in dry scrublands. There are about 38 species, or kinds, of mongooses. Read about one species, the suricate, on page 532.

Slender mongoose: 12 in (30 cm) long; tail, 10 in (25 cm)

▽ *Nipping and tumbling, two young dwarf mongooses play together. Dwarf mongooses cooperate when they care for young. One or two pack members stay with the offspring while others hunt.*

Dwarf mongoose: 8 in (20 cm) long; tail, 6 in (15 cm) when fully grown

Mongooses rest during the heat of the day. They sleep in termite mounds, in piles of rocks, among tangled tree roots, or in burrows.

MONGOOSE

LENGTH OF HEAD AND BODY: 8-25 in (20-64 cm); tail, 6-21 in (15-53 cm)

WEIGHT: 12 oz-11 lb (340 g-5 kg)

HABITAT AND RANGE: forests, open woodlands, grasslands, marshes, and scrublands throughout most of Africa, and parts of Europe, Asia, the Middle East, Madagascar, some Caribbean islands, and Hawaii

FOOD: small animals as well as fruit, nuts, and seeds

LIFE SPAN: 8 to 13 years in captivity, depending on species

REPRODUCTION: 1 to 6 young after a pregnancy of 2 or 3 months, depending on species

ORDER: carnivores

Some mongooses hunt during the morning and afternoon. Others move around at night.

To find food, a mongoose pokes its pointed nose into a hole, overturns rocks with its paws, or scratches in the dirt with sharp claws. Some find food by reaching underwater with their nimble fingers. If a marsh mongoose catches a hard-shelled animal such as a crab, it stands on its hind legs and hurls the crab to the ground to break it open and get at the meat. To crack an egg, the banded mongoose stands with its back to a large rock. It picks up the egg in its front paws and tosses it backward between its legs. The egg breaks against the rock.

A mongoose uses scent to communicate. It has several glands in its body that produce an oily

Banded mongoose: 15 in (38 cm) long; tail, 10 in (25 cm)

◁ *Banded mongoose crosses a round termite mound on a grassy plain near a lake in Uganda. Banded mongooses often roam in packs of 15 or more animals. During the day, they look for such insects as beetles. At night, the packs return to their dens, often in termite mounds.*

▽ *Tail held high, a ring-tailed mongoose looks for lizards. These rare mongooses live only on Madagascar, an island off the southeastern coast of Africa.*

Ring-tailed mongoose: 14 in (36 cm) long; tail, 9 in (23 cm)

substance. As a mongoose roams its home range, it sometimes leaves scent marks on certain objects along its trail. Dwarf mongooses and banded mongooses even mark one another. The scents help them recognize members of their group.

Some kinds of mongooses can have several litters a year. Each litter may include six offspring. Banded mongooses and dwarf mongooses cooperate among themselves in caring for young. One or two members of the pack clean, feed, and protect the offspring while the other adults hunt.

Yellow mongoose rests in the sun in South Africa. ▷
These animals live in family groups. They often share their burrows with other mongooses called suricates.

Yellow mongoose: 14 in (36 cm) long; tail, 10 in (25 cm)

375

Monkey

Japanese macaque: 21 in (53 cm) long; tail, 4 in (10 cm)

Snow dusts the thick fur of a female Japanese macaque and her young, huddled against the cold.

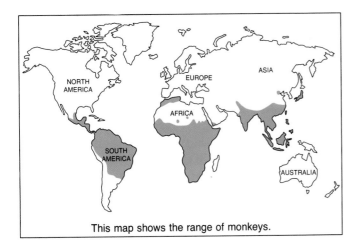
This map shows the range of monkeys.

Red spider monkey: 20 in (51 cm) long; tail, 27 in (69 cm)

WHAT'S MORE FUN than a barrel of monkeys? A whole forest full! Many of the world's forests are filled with monkeys—the familiar, intelligent animals that sometimes look and act like humans. That may be one reason why we find them so interesting. People usually think of monkeys as lively and playful. We say that children playing pranks are up to monkey business. If someone teases us, we often say, "Don't monkey around."

△ *Young red spider monkey gets a ride through a rain forest in Central America. To swing from branch to branch, its mother uses her tail like an extra arm.*

▽ *Three young chacma baboons, escorted by an adult, cross an open grassland in Africa. Baboons spend most of their waking hours on the ground. For safety, they sleep high in trees or on rocky cliffs.*

Chacma baboon: 32 in (81 cm) long; tail, 23 in (58 cm)

Monkey

Monkeys, along with apes and human beings, belong to the primate order. Apes, though they look somewhat like monkeys, have no tails. They are usually larger than monkeys. Chimpanzees, gibbons, gorillas, and orangutans are all apes. Read about them under their own headings.

There are about 125 species, or kinds, of monkeys in the world. People usually imagine monkeys swinging through trees in tropical forests. Many monkeys of South America and southern North America do behave that way. But monkeys also live in other habitats and climates. In Africa, monkeys live on open grasslands, as well as in forests. They are found in the mountains of Japan and on the tree-less plains high in the Himalayas of Asia. One kind makes its home in Europe. No monkeys, however, live on the continents of Australia or Antarctica.

Scientists place monkeys in two groups. North and South American monkeys are called New World monkeys. Those found in Africa, Asia, and Europe are called Old World monkeys.

Both groups of monkeys look and act much alike. They eat many of the same foods—fruit, nuts, seeds, leaves, flowers, insects, birds' eggs, spiders,

Pygmy marmoset bites into a grasshopper in a dense ▷ *Amazon rain forest. The animals—smallest of all monkeys—have speckled brown-and-yellow coats.*

▽ *Among the green leaves of an acacia tree in Kenya, a troop of olive baboons finds shelter.*

Olive baboon: 27 in (69 cm) long; tail, 20 in (51 cm)

Pygmy marmoset: 5 in (13 cm) long; tail, 8 in (20 cm)

△ Loud cries fill a forest in Venezuela as two male red howler monkeys announce their presence. Before they set out each morning to find food, they arouse animals in the forest with their calls. By howling, they warn other howler monkeys, "This is our territory. Keep out!"

NEW WORLD MONKEYS—*identified by their round, wide-set nostrils—make their homes in North and South America. Some New World monkeys have prehensile, or grasping, tails. Above, a red spider monkey uses its long tail to dangle from a tree.*

▽ *In Peru, a dusky titi gazes skyward. These small, red-bearded monkeys eat fruit, leaves, and flowers.*

and sometimes small mammals. But the bodies of monkeys in each group differ in certain ways. The nostrils of most New World monkeys are round and set far apart on short snouts. The nostrils of many Old World monkeys are curved and set close together on their snouts. Old World monkeys have pads of tough skin on their rumps. The pads cushion them while they sit to eat or to sleep. New World

Dusky titi: 15 in (38 cm) long; tail, 16 in (41 cm)

◁ *Upside-down weeper capuchin quenches its thirst at a river in Venezuela. These monkeys sometimes climb onto branches hanging just inches above the water. They lean down to take a drink.*

Weeper capuchin: 17 in (43 cm) long; tail, 17 in (43 cm)

▽ *Softly furred night monkey, or douroucouli, grips a tree in an Amazon rain forest. Sometimes called owl monkeys because of their large, round eyes, these South American primates sleep during the day. At night, they look for fruit, leaves, and flowers to eat.*

Night monkey: 13 in (33 cm) long; tail, 13 in (33 cm)

White-faced saki: 19 in (48 cm) long; tail, 16 in (41 cm)

△ *Vivid mask of light fur on its face identifies a white-faced saki as a male. Female sakis have only narrow, light-colored bands streaking their dark faces. Sakis live in the rain forests of South America.*

monkeys do not have this kind of built-in pillow.

Some Old World monkeys have pouches in their cheeks. These pouches allow them to store an extra supply of food to eat later. No New World monkeys have cheek pouches. But some do have a useful characteristic that no Old World monkey has: a prehensile (say pree-HEN-sul) tail. This tail can grasp objects and support the animal's weight.

Wrapping the tail firmly around a branch, a monkey can dangle upside down while eating. And it can use its tail to help it swing from limb to limb.

Some monkeys have hands like humans. These help them grasp while they are feeding and while they are traveling about. Although a few monkeys have no thumbs at all, many others have thumbs that help them hold on to objects. The guenons (say

Monkey

Like a guided missile, a squirrel monkey makes a ▷ spectacular leap from tree to tree. Its outstretched tail helps this South American monkey keep its balance.

▽ Bareback riders, young squirrel monkeys travel through a forest in Colombia with two older females. A female relative often cares for another's young.

Squirrel monkey: 12 in (30 cm) long; tail, 16 in (41 cm)

guh-NOHNS) of Africa and the capuchins (say kuh-PYOO-shunz) of Central and South America can move their thumbs around to touch some of their other fingers. They can pick up bits of food with their thumbs and fingertips. And their thumbs help them grip branches as they climb. Some New World monkeys cannot use their thumbs in this way. These monkeys must hold objects by pressing them between their fingers and the palms of their hands.

A few kinds of monkeys, such as baboons and some macaques (say muh-KACKS), spend much of the time on the ground. But most monkeys live in trees. Their slender, light bodies are well suited to climbing, swinging, and leaping. With their long, muscular arms and legs, they can move easily

Black-mantle tamarin: 9 in (23 cm) long; tail, 13 in (33 cm)

Cotton-top tamarin: 9 in (23 cm) long; tail, 15 in (38 cm)

△ White patch around its mouth looks like a broad smile on a black-mantle tamarin. To climb, tamarins and their close relatives, marmosets, use their long claws.
◁ Long mustache lends a dignified air to an emperor tamarin in Peru. This small monkey feeds on fruit and insects.

Black-and-white tassel-ear marmoset: 8 in (20 cm) long; tail, 13 in (33 cm)

△ Bushy white fur on its head gives the cotton-top tamarin its name.
◁ Forked tree provides a perch for a black-and-white tassel-ear marmoset. Silky tufts of fur hide the ears of this small monkey with the long name.

Emperor tamarin: 9 in (23 cm) long; tail, 14 in (36 cm)

383

Woolly monkey: 20 in (51 cm) long; tail, 25 in (64 cm)

△ *Young woolly monkey in Brazil reaches for a leaf. People often keep these fuzzy animals as pets.*
◁ *Gripping with toes, tail, and fingers, a red howler monkey plays acrobat in a rain forest in Colombia. Huge trees there provide howlers with a plentiful supply of a favorite food—fresh green leaves.*

among the trees. The star acrobat among monkeys is the spider monkey of Central and South America. It is named for its long, thin, spiderlike limbs and prehensile tail. It uses these to climb nimbly among the tallest trees in the forest. It makes its home in the topmost branches. There, as high as 100 feet (30 m) above the ground, it eats, sleeps, and raises its young. The spider monkey descends to the lower branches to feed on fruit and nuts.

Most monkeys are active during the day. As they climb, swing, and jump, they chatter, shriek, and scold one another. Only the night monkey, or douroucouli (say dur-uh-coo-lee), of South America moves around after dark. These monkeys usually travel in pairs.

MONKEY

LENGTH OF HEAD AND BODY: 5-40 in (13-102 cm); tail, as long as 36 in (91 cm)

WEIGHT: 4 oz-100 lb (113 g-45 kg)

HABITAT AND RANGE: forests, grasslands, and mountains in Africa, Asia, Europe, and North and South America

FOOD: leaves, fruit, insects, nuts, seeds, grasses, roots, birds' eggs, spiders, and small mammals

LIFE SPAN: as long as 45 years in captivity, depending on species

REPRODUCTION: 1 or 2 young after a pregnancy of about 4½ to 7½ months, depending on species

ORDER: primates

Probably the loudest voices belong to male howler monkeys, the largest of the New World monkeys. Each morning, their cries echo through the forest. They roar to let other monkeys know where they are. By their roars, howlers remind other howlers to keep out of their territories, or areas.

High in the trees of dense rain forests in South America, monkeys have few enemies other than eagles and hawks. The monkeys' most dangerous enemies are people who hunt them for food.

One monkey that has been hunted often in South America is the uakari (say wah-KAR-ee). When alarmed, the rare red uakari fluffs up its shaggy fur, and its face turns bright red.

Like humans, most monkeys usually have only one offspring at a time. The helpless newborn depends on its mother for warmth, food, and transportation. During the first days or weeks of its life, it

▽ *Red uakari, known by its scarlet face and bald head, pauses as it climbs up a slender tree trunk. Unlike most other monkeys, the rare uakari has a short tail.*

Red uakari: 18 in (46 cm) long; tail, 7 in (18 cm)

Monkey

OLD WORLD MONKEYS, found mostly in Africa and Asia, generally grow larger than their New World relatives. Curved and close-set nostrils and hard pads on their rumps help identify them. No Old World monkey has a prehensile tail.

Barbary macaque: 23 in (58 cm) long

△ *Barbary macaques rest in the sun. The two on the right groom each other near a mother and her offspring. The monkeys are often called Barbary apes because, like apes, they have no tails. Barbary macaques live in North Africa and on Gibraltar, a British territory at the southern tip of Spain.*

Bright white band on its upper lip gives a mustached guenon its name. Tropical ▷ forests of Africa provide homes for these fruit and leaf eaters. The animals sometimes sniff the ground looking for plants to eat. Special cheek pouches allow them to store food and to carry it back to the trees.

Mustached guenon: 19 in (48 cm) long; tail, 30 in (76 cm)

clings to the fur on its mother's belly as she moves about looking for food. Later the young monkey becomes more daring and playful. It rides on its mother's back or scampers about on its own. Sometimes it climbs onto the backs of other members of the monkey group. The adults are usually very patient with the young monkey.

Among some kinds of monkeys, the care of young is shared by several family members. Male marmosets (say MAR-muh-sets) and tamarins (say TAM-uh-rins) of South America carry their young, usually twins, on their backs. They may hand the offspring to the mothers only at feeding time.

Monkeys are social animals—that means they live in groups. While resting, they spend a great deal of time grooming other members of their group. One monkey combs through another's hair, picking out dirt and insects. Such careful grooming helps keep relations friendly among monkeys.

Groups of monkeys may vary in size. Some

△ *Rhesus monkey and her offspring eat yellow blossoms on the grounds of a Hindu temple in Nepal.*
◁ *Two hanuman langurs in India use bamboo stalks as stools. Like rhesus monkeys, they run free in villages and temples in Asia. Hindus consider monkeys sacred.*
▽ *Pig-tailed macaque pauses in a field in Indonesia. These large monkeys often raid crops for food.*

Pig-tailed macaque: 21 in (53 cm) long; tail, 8 in (20 cm)

Hanuman langur: 24 in (61 cm) long; tail, 31 in (79 cm)

387

Monkey

Lion-tailed macaque: 21 in (53 cm) long; tail, 13 in (33 cm)

Proboscis monkey: 25 in (64 cm) long;
tail, 26 in (66 cm)

△ How did the proboscis monkey of Borneo get its name? Proboscis means long, flexible snout. A male proboscis monkey's snout serves as a loudspeaker when he calls out to other monkeys in his group.

◁ Face framed by a shaggy mane, a lion-tailed macaque feeds on plants in India.

Black-and-white colobus monkey: 22 in (56 cm) long; tail, 32 in (81 cm)

Female black-and-white colobus monkeys in Kenya gather round as a mother (second from left) cradles her offspring. The tiny newborn still wears a coat of white. Adult monkeys often groom another's young.

monkeys live in pairs. They may stay with their mates for life, raising one offspring after another. Others live in larger troops, groups made up of adult males and females with young. Some troops contain one male and many females. In certain troops, a strong male acts as leader and protects the others.

Baboons, the largest of all monkeys, live and travel in tightly organized troops. Sometimes baboon troops number several hundred members. The hamadryas (say ham-uh-DRY-uhs) baboons of eastern Africa live in smaller groups. One male usually leads several females and their offspring.

Baboons are ground dwellers. These primates may roam as many as 12 miles (19 km) a day to look for food. They search for plants, roots, insects, and small mammals in the rocky, open countryside of Africa. Because baboons spend most of their time on the ground, they are in danger from leopards, cheetahs, wild dogs, and other hunters. If a baboon spies an enemy, it may try to run away. Or it may bark loudly and bare its sharp teeth. If necessary, it will fight to defend the troop. A male baboon, which may measure more than 3 feet (91 cm) long and weigh about 100 pounds (45 kg), is a fierce foe!

Monkeys range in size from the large baboon to the tiny pygmy marmoset of South America—the smallest monkey in the world. Its body is only 5 inches (13 cm) long—shorter than a toothbrush.

Monkeys differ in appearance and special abilities as much as they do in size. The short-haired woolly monkey of South America has a big potbelly. In fact, the Portuguese name for this large monkey, *barrigudo*, means "big belly." These animals eat huge quantities of fruit and leaves.

The Barbary macaque, also known as the Barbary ape, has no tail. But it too is a monkey. Barbary macaques live on the Barbary Coast of North Africa and on Gibraltar, a British territory at the southern tip of Spain. They are the only primates in Europe living outside of zoos. According to a long-held tradition, as long as the Barbary macaques remain on Gibraltar, the British will keep control of the area.

Many other kinds of macaques live in Africa

Young vervet clings to its mother in a forest in ▷ *southern Africa. A vervet's tail may measure as long as 3 feet (91 cm), almost twice the length of its body.*

Vervet: 21 in (53 cm) long; tail, 36 in (91 cm)

Monkey

△ *Two male olive baboons challenge each other on a grassland in Kenya. Fights like this sometimes occur among members of baboon troops. Such contests end quickly, however, when*

◁ *Baring long, sharp teeth, a male olive baboon threatens an unwelcome visitor. By this gesture, known as a threat-yawn, a baboon signals intruders to stay away.*

and Asia. The Japanese macaque, called the snow monkey, lives farther north than any other monkey. Its long, shaggy hair protects it from the cold. The rhesus (say REE-suss) monkey of India is also a macaque. It has been used in many experiments in medical research and in the exploration of space.

A large macaque, the pig-tailed macaque of Malaysia, helps out when inhabitants harvest coconuts there. Malaysians teach intelligent young animals to climb coconut palms and pick the ripe fruit. The monkeys drop the coconuts down to their owners who are waiting below.

Crab-eating macaques that live along riverbanks in southeastern Asia are very good swimmers. These long-tailed monkeys usually look for fish and shellfish at the water's edge. They add these to their diet of fruit and insects.

The small, easily trained capuchins are among the best-known monkeys in the world. In the past, street musicians called organ-grinders dressed the capuchins in costumes. As an organ-grinder played an instrument called a hand organ, his monkey performed stunts and collected money from onlookers.

In contrast to the playful capuchins, some monkeys may seem idle. Leaf eaters like the Asian langur (say LAHNG-gur) and the African colobus (say KAHL-uh-bus) monkeys spend many hours each day just sitting and digesting their food. But when they do move, these monkeys can leap amazing distances. Colobus monkeys have been seen jumping as far as 30 feet (9 m) from one tree to another.

In some parts of the world, monkeys are considered sacred. In India, langurs live on the grounds of Hindu temples. They often roam unharmed in towns and villages. People bring them offerings of food—bits of pumpkin or handfuls of rice. Langurs are familiar sights in open-air markets, where they help themselves to fruits and vegetables.

Hindus in India consider all monkeys to be holy. One kind of langur—the hanuman (say HUN-uh-mahn) langur—is named for the Hindu monkey god, Hanuman. According to Indian mythology, Hanuman bravely came to the aid of a prince. So, to Hindus, monkeys are a symbol of devotion.

one male gives in to the other. The strong males in a troop usually help each other protect the rest of the baboons from enemies.

△ Up on her hind legs, a young female olive baboon stretches to pick an acacia pod. Baboons also eat fruit, grasses, and small animals. They roam several miles a day in search of food.

▽ Members of a hamadryas baboon troop gather on a sunny rock. The leader, a large light-colored male, has a long cape of fur and full side-whiskers. While resting, baboons often groom one another's fur.

Hamadryas baboon: 27 in (69 cm) long; tail, 19 in (48 cm)

391

Moose

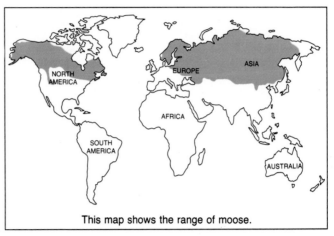

This map shows the range of moose.

MOOSE

HEIGHT: 5½-7 ft (168-213 cm) at the shoulder

WEIGHT: 500-1,800 lb (227-816 kg)

HABITAT AND RANGE: forests in northern North America, Europe, and Asia

FOOD: bark, leaves, twigs, shrubs, herbs, and underwater plants

LIFE SPAN: 15 to 20 years in the wild

REPRODUCTION: 1 or 2 young after a pregnancy of about 8 months

ORDER: artiodactyls

LAKES, STREAMS, AND MARSHES provide feeding grounds for moose in forests in North America, Europe, and Asia. The moose is the largest member of the deer family. But it does not look much like its smaller, more delicate relatives. The moose has a long face and a muzzle that hangs loosely over its chin. A large fold of skin, called a bell, dangles at the animal's throat. Its thick neck is topped with a heavy mane that stands up when the moose becomes angry or alarmed.

The bull, or male, moose of Alaska is the biggest moose of all. It measures 7 feet (213 cm) tall at the shoulder and may weigh as much as 1,800 pounds (816 kg). Its coat, dark brown in summer, turns a lighter gray-brown in winter.

The moose's legs are too long for easy grazing. The animal must kneel to reach the ground. Or, like

Antlers of two bull moose of Alaska stand out against a background of blazing fall colors in the foothills of Mount McKinley. Largest member of the deer family, the moose feeds on leaves and shrubs.

a giraffe, it must spread its forelegs wide apart and bend its head down. More often, the moose browses, feeding on shrubs, tree bark, and the leaves and twigs of higher branches.

A moose chews a mouthful only a few times before swallowing. Then it quickly takes another bite. After eating, the animal brings up a wad of partly digested food, called a cud. It chews the cud thoroughly, swallows, and completely digests it.

In summer, a moose often feeds in water. Wading chest-deep, the animal nibbles water lilies and other water plants. It often dips its head beneath the surface. With its sensitive lips, it feels for plants and tears them from the bottom.

Occasionally, the large animal feeds in deep water. Diving as far down as 18 feet (5 m), it can stay under the surface for half a minute at a time.

A moose walks easily over marshy ground on its large, split hooves. Broad feet help support the moose and keep it from sinking in the soft earth. In winter, a moose can move without difficulty through snowdrifts as deep as 3 feet (91 cm).

A moose has keen senses of smell and hearing. Its long ears turn to catch the faintest rustle. At the sound or scent of an enemy—usually a wolf or a bear—a moose stands motionless. If the enemy comes close, the moose races off.

Over a long distance, a moose is able to trot at about 20 miles (32 km) an hour. When threatened, the animal can run 35 miles (56 km) an hour in a

Moose

◁ *Long-legged twins explore a forest clearing while their mother feeds on a leafy shrub. Moose calves lack the spotted coats that help hide their deer relatives. But a cow moose fiercely protects her young.*

△ *Dripping wet, a cow moose pulls up a mouthful of food. With nostrils closed, a moose dips its head underwater to feed. Sensitive lips pluck plants from the bottom.*

◁ *Leaving a trail of spray, a young moose dashes through a lake in Alaska. When only five days old, a calf can outrun a person.*

sprint. Sometimes a moose tries to lose its pursuer by diving into a river or a lake. The moose is a strong swimmer. Paddling with its hooves, a moose can swim for several miles without stopping. The hollow hairs of its coat are filled with air. They help keep the animal afloat.

Except during the mating season, adult moose live separately. Bulls usually travel alone. Females, or cows, feed by themselves or with their calves.

Usually two reddish brown calves are born to each cow in spring. Calves lack the spotted coats of some other newborn in the deer family. Moose young do not rely on their coloring to hide them from enemies. Their mother fiercely guards them.

At first, calves drink their mother's milk. Soon they begin to eat plants as well. Weighing about 30 pounds (14 kg) at birth, each calf doubles its weight within a few weeks. For the first year, the young moose follow their mother. Then, shortly before she

△ *Antlers fully grown and hardened in early fall, a bull moose thrashes the bony growths against tree branches. The velvet that nourished the antlers peels off painlessly.*

gives birth again, she drives the year-old calves away. They learn to live on their own.

At the end of a male calf's first summer, two bony bumps begin to rise on his forehead. The next spring, a set of short spike antlers will form. Adult males grow new antlers in spring. They shed their old sets from December to February.

Strong and solid, moose antlers are flat at the center like the palm of a person's hand. Their sharp edges may have more than forty points. The largest antlers belong to moose in Alaska. They may measure more than 6 feet (183 cm) from tip to tip.

While growing, a bull's antlers are protected by a covering of soft skin and fine hairs called velvet. Blood vessels in the velvet nourish the bony growths. Tender and rubbery during spring and summer, the antlers are fully grown and hardened by early fall. The bull rubs his head against shrubs and trees, peeling the velvet off in strips. Shedding

the velvet looks bloody, but the process does not hurt the moose. A bull keeps his antlers for just a few months. He uses them only during the mating season, in September and October.

At that time, bulls have contests to find out which is the stronger. Frequently, two males clash head-on. The powerful animals brace their antlers and push and shove. Usually the weaker one turns away. But sometimes the moose fight to the death. The stronger bull mates with a cow and stays with her for several days. Then he finds another mate.

When the mating season ends, the bulls leave the cows. Adult males and females may not meet again for a year. But where food is plentiful, moose may gather to feed in the same area. Even when many moose are close together, cows and bulls pay no attention to each other.

Read about other members of the deer family—caribou, deer, and elk—in their own entries.

Searching for a mate, a bull moose (above) sniffs the air and listens carefully. Cows call to a bull with a moaning cry. Bulls answer with hoarse grunts. Cautious bulls (left) confront each other nose to nose. The older bull shows his larger antlers to the younger one. If the younger bull does not retreat, the two will battle by furiously pushing each other with their antlers. The winner will mate with a cow moose.

Mouflon

The mouflon is a kind of wild sheep. Read about sheep on page 506.

Mouse

CITIES AND FARMS, deserts and mountains, meadows and forests—mice live in all kinds of habitats almost everywhere in the world. These small rodents make their homes wherever they can find food and shelter. Because of their size—usually less than 6 inches (15 cm) long, not counting their tails—they can hide in small spaces.

Mice resemble their close relatives, rats, except that mice are generally smaller. Mice usually have pointed noses, rounded ears, and skinny tails. Most have soft gray or brown fur, but some mice have special coats or markings. Stiff, sharp hairs cover the backs of spiny mice of Africa and Asia. The striped

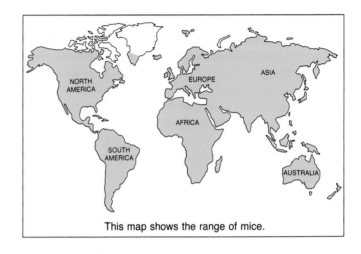

This map shows the range of mice.

Golden mouse: 3 in (8 cm) long; tail, 3 in (8 cm)

House mouse: 3 in (8 cm) long; tail, 3 in (8 cm)

△ *Female house mouse nurses her blind, hairless young. Every year, she may have five or more litters with as many as twelve young in each.*
◁ *Like a tiny singer, a southern grasshopper mouse in Mexico tilts back its head and squeaks to the sky. Its calls probably warn other mice to stay away.*

grass mouse of Africa has stripes on its back that help the animal hide in grassy surroundings.

Most mice scurry along the ground. They use their long whiskers to feel the way. Some kinds of mice, though, move by hopping on powerful hind legs. The woodland jumping mouse of North America is able to escape its enemies by bounding quickly away. It may jump as far as 7 feet (213 cm), using its long tail for balance.

Other mice scamper up and down tree trunks and along twigs and slender branches. The African

◁ *Golden mouse in Florida uses its forepaws to clean its whiskers. These mice usually live in trees and bushes. They can grasp branches and vines with their tails.*
▽ *Holding a hazelnut in its forepaws, a wood mouse feeds in front of its deep burrow. Wood mice live in parts of Europe, Asia, and Africa.*

Wood mouse: 4 in (10 cm) long; tail, 3 in (8 cm)

MOUSE

LENGTH OF HEAD AND BODY: 2-6 in (5-15 cm); tail, 1-8 in (3-20 cm)

WEIGHT: $1/_4$ oz-2 oz (7-57 g)

HABITAT AND RANGE: almost every type of habitat worldwide

FOOD: plants and some small animals

LIFE SPAN: as long as 6 years in captivity

REPRODUCTION: 1 to 12 young after a pregnancy of about 1 month

ORDER: rodents

Southern grasshopper mouse: 4 in (10 cm) long; tail, 2 in (5 cm)

Mouse

Some activities in the daily lives of European harvest mice: Below, two harvest mice greet each other in a wheat field in England. At right, a harvest mouse perches among stalks of wheat. It holds its tail in its tiny paws and grooms it.

climbing mouse often winds its long tail loosely around a twig to help it climb.

One species, or kind, of mouse is called the house mouse because it often lives in houses and other buildings. House mice like the food and warmth they find where people live. House mice make their nests in walls of houses, in cupboards, in boxes in attics, and even in pockets of coats hanging in closets. Mice have been found living in cold-storage rooms, burrowed in frozen meat.

Most other kinds of mice—and many house mice, too—build their nests away from people. The pencil-tailed tree mouse of Asia often lives inside a bamboo stalk. The Florida mouse may build its nest in a burrow made by a turtle. In the fall, the European harvest mouse weaves blades of tall grass together to form a ball-shaped nest.

The Australian field mouse usually digs a burrow on a sandy plain. It hides the entrances to its burrow by piling twigs and plants around the holes. The white-footed mouse of North America sometimes nests in a hole in a dead tree or moves into an unused bird's nest.

Mice not only live almost anywhere, but they also eat almost anything—including glue, leather, and soap. In the wild, mice nibble grain, roots, fruit, leaves, seeds, stems, and grasses. The pocket mouse of western North America carries seeds in fur-lined pouches on the sides of its face. When a pocket mouse is ready to empty these pouches, it pushes the food out by flicking its forepaws against its cheeks. A pocket mouse usually takes its food to a chamber in its burrow. There it stores the food for later use.

Some kinds of mice also eat meat. The grasshopper mouse, which lives in dry regions in western North America, feeds on grasshoppers, other insects, scorpions, worms—and even other grasshopper mice! This fierce little mouse sometimes stands on its hind legs, tosses back its head, and lets out a loud squeak—probably to warn other grasshopper mice to stay away.

When they do not live in houses, house mice may stay together in large family groups. All the mice in a group share a burrow. They recognize each other by scent. The mice take turns grooming one

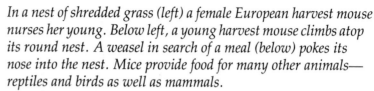

In a nest of shredded grass (left) a female European harvest mouse nurses her young. Below left, a young harvest mouse climbs atop its round nest. A weasel in search of a meal (below) pokes its nose into the nest. Mice provide food for many other animals—reptiles and birds as well as mammals.

another. If any mice from other groups try to come into the burrow, they are attacked and driven out.

In each group of house mice, one strong male is always the leader. But this mouse may be challenged by another male. The two mice bite and claw each other until one scurries away. If a burrow becomes too crowded, younger mice will leave and dig their own burrow.

Female mice have several litters of young every year. There may be as many as 12 tiny, helpless young in each litter. Most newborn mice weigh much less than 1 ounce (28 g). They may nurse for a month, but they begin eating solid food about two weeks after birth. Then they begin to follow their mother on short trips outside the burrow.

Almost all meat-eating animals, from weasels

▽ Black eyes shining, a harvest mouse eats a grain of wheat cradled in its forepaws. Harvest mice spend much of the day looking for seeds, grasses, and insects.

and cats to hawks and owls, prey on mice. Only one or two young in a litter will grow up to become adults. Some mice live only two or three months in the wild. But because they have young so often, many kinds are plentiful.

Mice are killed by humans, too. People try to get rid of mice because the small rodents can be pests. Mice eat or spoil millions of dollars' worth of grain and other foods every year. Some carry diseases. They can do damage just by gnawing on wood and clothing. Like all rodents, mice have front teeth that grow throughout their lives. They must gnaw to keep these teeth worn down.

Mice can sometimes be helpful in the fight against disease. Scientists use mice in experiments that may lead to cures for illnesses. Mice also play an important role in nature. They form one link in a food chain from plants to meat-eating animals. Mice feed mainly on plants. They themselves provide food for other animals—birds, reptiles, and mammals.

African climbing mouse: 3 in (8 cm) long; tail, 4 in (10 cm)

△ *Holding on to a stem with its tail, an African climbing mouse in Tanzania works its way toward the ground. These mice look for food in trees and on vines.*

Deer mouse: 3 in (8 cm) long; tail, 3 in (8 cm)

Tasmanian mouse: 4 in (10 cm) long; tail, 4 in (10 cm)

Australian hopping mouse: 4 in (10 cm) long; tail, 5 in (13 cm)

Mice can move their tails into many positions. A deer mouse (top) curls its tail under its body as it sits near its nest in a fallen tree. A Tasmanian mouse (above) wraps its tail over a branch as it rests while searching for food. For balance, an Australian hopping mouse (left) holds its tail out behind as it moves across the ground.

Long, coarse coat of a male musk-ox sweeps the ground as he travels over ice-crusted snow.

Musk-ox

(*say* MUSK-OX)

FOR THOUSANDS OF YEARS, shaggy, humped musk-oxen have lived in cold, rocky areas of the arctic region. Musk-oxen are well equipped for life on the tundra, a harsh, treeless land. They have thick coats that protect them during the cold winters. An outer layer of long hairs, called guard hairs, covers a shorter undercoat. The outer coat hangs down nearly to the ground. When warmer weather comes in spring, the soft, woolly undercoat begins to fall out through the long guard hairs.

A musk-ox's curved hooves help the animal travel easily across the rugged arctic plain. Its hooves have sharp rims and soft pads. They allow the musk-ox to paw through the snow for food and to clamber over rocky slopes.

Musk-oxen travel in herds during the entire year. A female, called a cow, sometimes leads the herd. The animals spend the warmer months

This map shows the range of musk-oxen.

MUSK-OX

HEIGHT: 4-5 ft (122-152 cm) at the shoulder

WEIGHT: 500-800 lb (227-363 kg)

HABITAT AND RANGE: tundra in Alaska, northern Canada, Greenland, Norway, Siberia, and some arctic islands

FOOD: grasses, willows, and some arctic flowers

LIFE SPAN: 12 to 20 years in the wild

REPRODUCTION: 1 young after a pregnancy of about 8 months

ORDER: artiodactyls

403

feeding near streams and lakes. There they eat grasses, willows, and some flowers.

Traveling in a herd offers protection against enemies. When a wolf approaches, musk-oxen defend themselves by forming a line. If the wolf comes even closer, the adults take turns driving it away.

When threatened, both male and female musk-oxen are fierce fighters. Their sharp horns look like huge, hard bows. They curve down beside the animals' faces, then turn up at the ends. The horns of a male, called a bull, form a bony shield across his forehead. A musk-ox charges with its head lowered. It tries to gore its enemy with the tips of its horns.

When the summer mating season comes, bulls may begin to fight among themselves for females. The stronger bulls try to drive the weaker ones away by charging at them. Again and again, the bulls crash together, horn to horn, until one bull gives up.

Calves are born after a pregnancy of about eight months. A newborn can keep up with its mother

only a few hours after birth. It grows quickly. By the time it is five or six years old, a musk-ox has reached its full size. A bull may weigh as much as 800 pounds (363 kg) and may measure almost 5 feet (152 cm) tall at the shoulder. Cows are smaller and lighter.

For many years, musk-oxen were hunted for their hides and meat. So many were killed that only a small number remained. Today there are laws to protect musk-oxen. Herds have been put on special preserves in Alaska, Norway, and Siberia.

△ *Ready for attackers, a herd of musk-oxen (above) forms a line for defense. When surrounded, the animals often make a protective circle. Huge horns frame the face of an old male musk-ox (top). Patches of fur under his eye show bits of his winter coat. As the weather becomes warmer, he sheds this shorter, woolly undercoat.*

◁ *Small herd of shaggy, brown musk-oxen pounds across the beach of Nunivak Island, off the coast of western Alaska. Herds live there on a preserve.*

Muskrat nibbles on a bulrush in shallow water in Montana. Muskrats feed on plants and shellfish.

Muskrat

(*say* MUSK-rat)

LIKE THE BEAVER, the muskrat often builds a dome-shaped home in the water. This rodent piles up water plants in a marsh until a mound is formed. Mud helps hold the muskrat's mound together. The animal then dives to the base of the mound. It gnaws out tunnels as well as a living space inside, above the water level. There it is safe from enemies such as turtles and alligators.

Other muskrats make their homes in burrows that they dig in the banks of streams or on the shores of lakes. These muskrats dig tunnels that open underwater. Then muskrats can swim and hunt for food even when ice covers the surface.

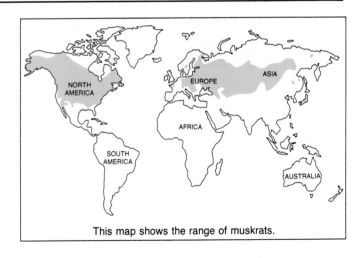

This map shows the range of muskrats.

406

Muskrats live in North America. People have also taken them to Europe and Asia, where they now live in the wild. Muskrats are excellent swimmers and divers. They paddle with their partly webbed hind feet, and they steer with their scaly tails. Muskrats eat mainly water plants, but sometimes they feed on shellfish.

A female muskrat usually bears five to seven blind, hairless young twice a year. The young begin to swim within their first three weeks.

On land, muskrats move awkwardly. There they are hunted by foxes and raccoons. Trappers also take muskrats for their thick brown fur. The animals have scent glands that produce a strong-smelling substance. Its musky odor helps explain how the animal got its name.

On the frozen surface of a lake in Wyoming, a muskrat ▷
looks for food in early winter.
▽ *Muskrat house rises in a marsh in North Dakota. Muskrats pile up grasses to form mounds. By cutting down these plants, they sometimes create moats around their homes.*

MUSKRAT

LENGTH OF HEAD AND BODY: 10-14 in (25-36 cm); tail, 8-11 in (20-28 cm)

WEIGHT: 2-4 lb (1-2 kg)

HABITAT AND RANGE: marshes, lakes, streams, and ponds in parts of North America, Europe, and Asia

FOOD: mainly water plants and shellfish

LIFE SPAN: about 4 years in the wild

REPRODUCTION: usually 5 to 7 young after a pregnancy of about 1 month

ORDER: rodents

Narwhal
The narwhal is a kind of porpoise. Find out about porpoises on page 452.

Numbat
(say NUM-bat)

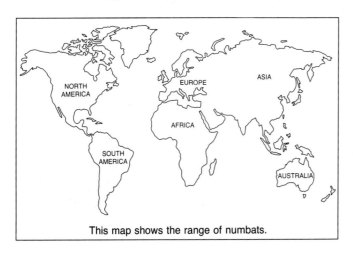
This map shows the range of numbats.

NUMBAT

LENGTH OF HEAD AND BODY: 7-11 in (18-28 cm); tail, 5-7 in (13-18 cm)

WEIGHT: 10-16 oz (284-454 g)

HABITAT AND RANGE: forests and plains of southwestern Australia

FOOD: termites and some ants

LIFE SPAN: about 5 years in captivity

REPRODUCTION: 2 to 4 young after a pregnancy of about 1 month

ORDER: marsupials

FROM SUNRISE TO SUNSET, the nimble numbat hunts for termites, its favorite food. It climbs dead trees, searching for the wood-eating insects. It scampers over old stumps and fallen limbs. Sharp senses of sight and smell help this animal find its prey. A numbat can eat several thousand termites a day! It also eats ants that invade termite nests.

To get at the insects, a numbat uses its sharp front claws. It scratches the ground or rips open rotten logs. Then its long, thin tongue darts out and laps up termites and ants.

At sunset, the numbat returns to its den in a shallow hole or a hollow log. There it sleeps in a nest of dry leaves and grass. Except during the mating season, numbats live alone.

Most numbats live in the eucalyptus forests of southwestern Australia. Their grayish brown coats blend well with the dead trees and dried leaves of the forest floor. Some numbats live on the plains just east of the forests. They have brick-red fur. All the animals have bushy tails and bold white bands across their backs. These markings give the numbat another name: banded anteater.

Once a year, a female numbat gives birth to as many as four tiny offspring. Each newborn climbs onto its mother's belly and attaches itself to a nipple. It will stay there for four months as it grows.

The numbat is a marsupial (say mar-soo-pea-ul). Most marsupials have pouches for their young. But numbats have no pouches. Long fur on the female numbat's belly protects her offspring. Many other Australian mammals are marsupials. Read about bandicoots, kangaroos, phalangers, and wombats under their own headings.

Bushy tail held high, a numbat follows its sensitive ▷ *nose to prey. With sharp claws, the animal rips open rotten wood where termites live. The numbat uses its quick, darting tongue to lap up the insects. Unlike other marsupials, numbats move about in the daytime.*

Nutria
Nutria is another name for the coypu. Read about the coypu on page 166.

Nyala
The nyala is a kind of antelope. Learn about antelopes on page 52.

Ocelot
The ocelot is a kind of cat. Read about cats on page 126.

Okapi

(say oh-COP-ee)

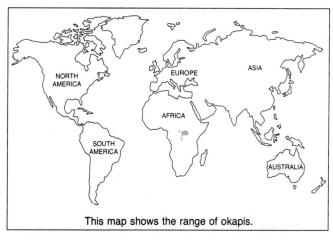

This map shows the range of okapis.

Dazzling leg stripes mark the chocolate-colored coat of an okapi in Africa. Scientists began to learn about this shy forest dweller less than a century ago.

A CENTURY AGO, scientists did not know that the okapi even existed. Then, in the 1890s, European explorers obtained skins of the animal from tribesmen who hunted it deep in the rain forests of central Africa. News reports of this shy, dark-coated mammal—a relative of the giraffe—made newspaper headlines all over the world.

Okapis have brown bodies with creamy-white stripes on their legs and hindquarters. The stripes help hide the horse-size animals in the forest. Both males and females have long, muscular necks. Like giraffes, they have short, skin-covered knobs on their foreheads. The forehead knobs of males are bigger than those of females.

Okapis eat leaves, which they pick with their long, gray tongues. The animal's tongue is so long

that it can reach up and lick the corners of its eyes as it washes its face.

Little is known about the behavior of okapis in the wild because only a few scientists have observed them there. Several animals may live in the same area, but they do not travel together in herds. Except for mothers with young, each adult roams alone, following paths worn smooth by other okapis.

A female okapi gives birth to a single young. Only half an hour after its birth, the infant can stand. It begins eating leaves after a few weeks.

OKAPI

HEIGHT: 59-67 in (150-170 cm) at the shoulder

WEIGHT: 440-550 lb (200-249 kg)

HABITAT AND RANGE: rain forests of central Africa

FOOD: buds and leaves of shrubs and trees

LIFE SPAN: about 25 years in captivity

REPRODUCTION: usually 1 young after a pregnancy of about 15 months

ORDER: artiodactyls

Opossum

(*say* uh-PAHS-um)

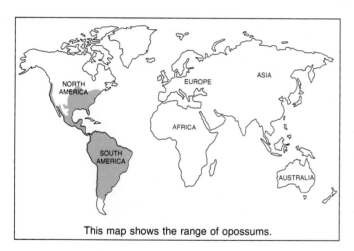

This map shows the range of opossums.

HAVE YOU EVER "PLAYED POSSUM"? Have you kept still and pretended to be asleep when you really weren't? This trick gets its name from the way an opossum sometimes acts. Playing possum can save its life. When an opossum is attacked by an enemy such as a fox or a dog, it lies so still that it seems to be dead. The opossum's eyes close, or the animal stares without blinking. Its tongue hangs out. The attacker may shake the opossum or toss it about. Even then the victim doesn't move.

Most attackers soon lose interest in what seems

Hanging on with all four feet, a young Virginia ▷
opossum looks down from a branch of a wild cherry tree.
Enemies such as foxes and dogs cannot reach it there.

Virginia opossum: 17 in (43 cm) long; tail, 13 in (33 cm)

to be a dead animal. They leave to look for livelier prey. Cautiously, the opossum lifts its head and looks about. If the danger is gone, it scurries off.

There are more than 65 kinds of opossums. They live in wooded areas in many parts of North

◁ *Hollow tree in New Jersey makes a cozy den for a Virginia opossum. Opossums in North America also make their homes in the empty dens of such large rodents as marmots. The opossums line their nests with leaves.*

▽ *Nearly too big to ride piggyback, three-month-old Virginia opossums still scramble aboard their mother. Soon they will go off to live and hunt by themselves. Within a year, they may have offspring of their own.*

Coiled around a twig, the long tail of a mouse ▷ opossum helps the animal keep its balance. This kind of opossum has no pouch. As many as 15 tiny offspring simply hang on to the nipples on their mother's belly.

Mouse opossum: 3 in (8 cm) long; tail, 4 in (10 cm)

Gray four-eyed opossum: 13 in (33 cm) long; tail, 11 in (28 cm)

△ *Gray four-eyed opossum of South America sniffs for food on a tree limb. The animal has only two eyes. It gets its name because of the white spots on its forehead.*

and South America. But only one kind plays possum—the Virginia, or common, opossum of North America. This opossum is the largest of all—about 2½ feet (76 cm) from nose to end of tail. It is also the only kind found in the United States.

Millions of years ago, opossums probably lived only as far north as southern North America. Gradually, they spread farther. By the 1600s, their range included what is now Virginia. Today the animals are also found in Canada.

Opossums are marsupials (say mar-soo-peaulz), or pouched mammals. They give birth to tiny, underdeveloped young. Just after birth, the young crawl into their mother's pouch, where they feed and continue to grow. Opossums are the only marsupials that live in North America. Most marsupials, like kangaroos and koalas, live in Australia and on its neighboring islands. Read about kangaroos on page 310 and koalas on page 318.

Virginia opossums have one or two litters a year. As many as twenty young are born at one time. But usually no more than eight offspring survive. Newborn opossums are blind and hairless. No bigger than bees, they have tiny claws that help them crawl into the pouch.

The young stay in the pouch for about two months. Then they begin to go in and out. They nurse for another month. When their mother hunts for food, they cling to the fur on her back.

Newborn mouse opossums ride with their mother even though she has no pouch. At first, the young dangle from her nipples. They are protected only by her fur. If a tiny opossum falls off, it makes a high-pitched sound. Human beings cannot hear its call, but its mother can. She goes back to find it. After a month, young mouse opossums crawl onto their mother's back and ride there. Mouse opossums are among the smallest opossums. They are found from Mexico south into Argentina.

When opossums reach full size, they leave their mother and go off to live alone. Some kinds have no regular shelters. They may sleep in a new place every day. Virginia opossums sometimes use old marmot burrows. Or they may make their dens in hollow trees. The opossum lines its den with leaves. It hooks its tail around a bunch of leaves and carries them back to its den.

The opossums that live in tropical forests—woolly opossums, four-eyed opossums, and mouse opossums—spend most of their time in the trees. There they often build nests of twigs and leaves.

Opossum

◁ Hanging upside down, a woolly opossum licks nectar from night-blooming flowers in Panama. Woolly opossums have thick fur and tails longer than their bodies. Their tails help them hold on to branches.

Sometimes a mouse opossum simply moves into an empty bird's nest.

Climbing trees is easy for an opossum. Its sharp claws hook into the bark. Its rear paws help it hold on to small branches. Its long tail can also grip branches, as if it were an extra hand. This helps an opossum keep its balance as it travels about.

Opossums sleep and rest most of the day. At night, they search for food such as grasses, fruit, nuts, insects, worms, and even snakes. They eat mice and birds if they can catch them. They also eat dead animals. Opossums will eat almost anything they can find. Those that live near people—on farms, in suburbs, and even in city parks—will go after chickens or search for garbage.

If an opossum has been prowling around a house, its tracks are easy to spot. An opossum's tail leaves a wavy mark in the dirt. On either side of this mark are paw prints. Because an opossum's big toes are far apart from the other toes on its hind feet, its tracks look like little human handprints.

Other marsupials called possums live in Australia. They are members of the phalanger family. Read about them on page 436.

◁ With big brown eyes open wide, a woolly opossum peers into the shadows of a tropical forest in Peru. A dark stripe runs between the animal's eyes, from the top of its head almost to the tip of its nose. This nimble climber hunts for fruit and insects in the trees. It moves around in the dark and rarely comes down to the ground.

Woolly opossum: 10 in (25 cm) long; tail, 13 in (33 cm)

OPOSSUM

LENGTH OF HEAD AND BODY: **3-20 in (8-51 cm); tail, 4-21 in (10-53 cm)**

WEIGHT: **$1/_2$ oz-12 lb (14 g-5 kg)**

HABITAT AND RANGE: **wooded areas in parts of North and South America**

FOOD: **grasses, fruit, nuts, insects, reptiles, and other small animals**

LIFE SPAN: **about 3 years in the wild**

REPRODUCTION: **as many as 20 young after a pregnancy of about 2 weeks**

ORDER: **marsupials**

Orangutan

(*say* uh-RANG-uh-tan)

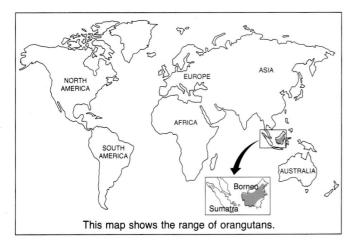

This map shows the range of orangutans.

ORANGUTAN

LENGTH OF HEAD AND BODY: 44-54 in (112-137 cm)

WEIGHT: 73-180 lb (33-82 kg)

HABITAT AND RANGE: tropical rain forests in parts of Sumatra and Borneo

FOOD: mostly fruit and leaves and some flowers, bark, and insects

LIFE SPAN: more than 50 years in captivity

REPRODUCTION: usually 1 young after a pregnancy of about 8½ months

ORDER: primates

Stretching out its lower lip, a young orangutan uses a leaf to get a drink. The animal dips the leaf into a puddle and lets the water run into its mouth. These orange apes live only on the islands of Sumatra and Borneo.

"PERSON OF THE FOREST"—that's the meaning of the word *orangutan* in the Malay language of Southeast Asia. These long-haired orange apes live in tropical rain forests in parts of Sumatra and Borneo. Like the other apes—chimpanzees, gorillas, and gibbons—orangutans belong to the primate order, which also includes human beings.

For the size of an orangutan's body, its arms are extremely long. Stretched out to the sides, the arms of a full-grown male may measure 7 feet (213 cm) from fingertip to fingertip. When the animal stands upright, its hands reach almost to the ground. An orangutan uses its strong arms to climb and to move among the trees, where it spends much of the time.

Adult males may grow to be about 4½ feet (137 cm) long. They weigh about 155 pounds (70 kg). Females weigh about half that much. As they become older, male orangutans develop big cheek pads on their faces and pouches on their throats. The females seem to prefer bigger males as mates.

The orangutan is a less social animal than other apes. Most of the time, an adult male travels alone through the forest. He lets other orangutans know of his presence by giving a "long call"—a series of grumbling and burbling sounds ending in a bellow. This warns other males to stay away—but females may be attracted by the call.

415

Orangutan

When males and females mate, they stay together for several days. About eight and a half months after mating, a female orangutan gives birth, usually to one young. The newborn is almost entirely helpless. Although the infant can hold on to its mother soon after birth, she usually cradles it in her arms.

A young orangutan depends on its mother for warmth, food, and transportation for a long period.

◁ *Covering its forehead with its hands, an orangutan takes a rest. Orangutans begin searching for food early in the morning. At midday, they nap.*

Using long, powerful arms to pull herself up, a ▷ female orangutan scales a tall tree on Borneo. Orangutans spend more time in the trees than do their African relatives—chimpanzees and gorillas. As loggers and farmers cut down more of the forests, orangutans have fewer places to roam and to find food.

▽ *Hefty adult male orangutan huddles in a nest in the trees. The animal bends leafy branches down to make a safe place to sleep. Every night, he builds a new nest.*

Orangutan

Natural rain bonnet protects a young orangutan ▷
from wet weather on Sumatra. During downpours,
orangutans often take shelter under leafy covers.

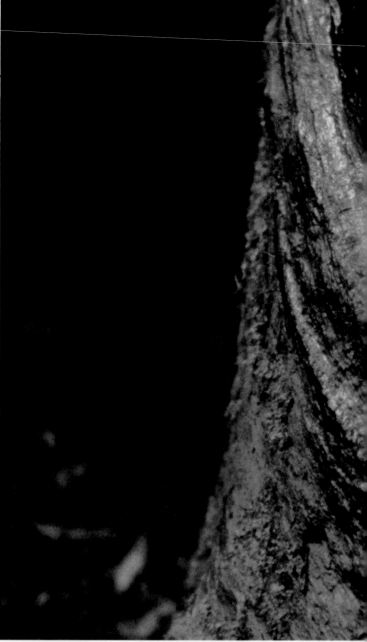

△ Twisting and
stretching, a young
female orangutan
practices the climbing
skills she needs for life in
the rain forest.
Still clinging to its ▷
mother after two years,
an orangutan rides on the
adult's back. Mothers
and young usually stay
together for six or seven
years, until the offspring
can take care of itself.

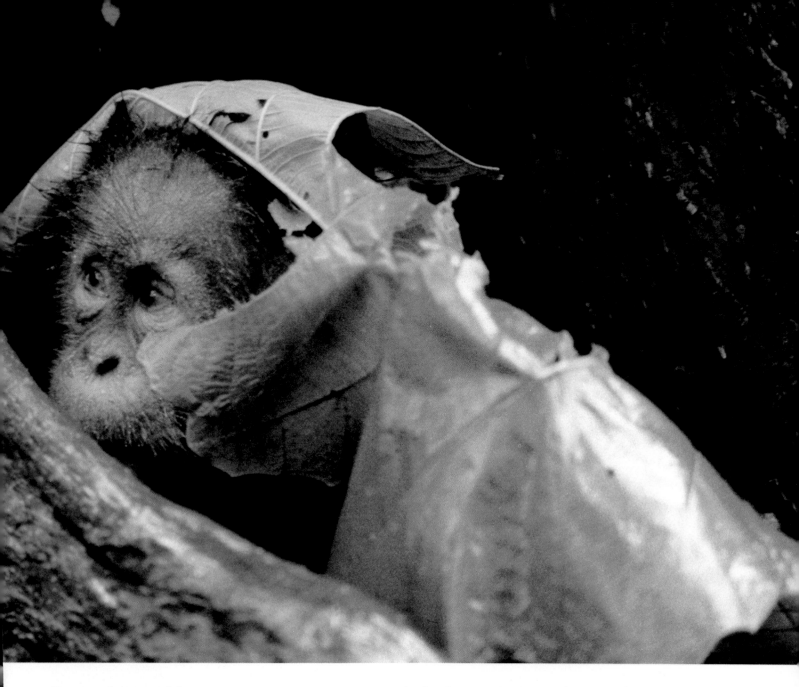

Even at the age of four years, a young orangutan does not stray far from its mother. It is just beginning to climb and to search for its own food. An orangutan will not become independent until it is six or seven years old.

Orangutans eat mostly fruit and leaves. They also feed on flowers, bark, and occasionally on insects. They rise early in the morning and spend the day looking for food. They may travel as far as 1 mile (2 km) a day to reach trees that have fruit.

Before dark, orangutans make sleeping nests by bending down branches in the trees. Orangutans often build day nests for naps. They sometimes put together an overhead cover of leaves for shelter during heavy rainstorms. Scientists have even seen orangutans hold leafy branches over their heads to protect themselves during downpours—just as people use umbrellas.

Orangutans live for a relatively long time. In captivity, some have reached more than fifty years of age. These apes have few enemies. The greatest danger they face comes from humans. People capture young orangutans to sell as pets. And as loggers and farmers cut down the forests where orangutans live, their numbers become smaller. Laws have been enacted and reserves have been set aside to help protect the animals. But orangutans are still among the most endangered of all primates.

Oryx

The oryx is a kind of antelope. Read about antelopes on page 52.

Otter

(say OTT-er)

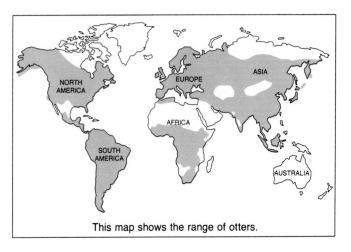

This map shows the range of otters.

OTTERS HARDLY EVER REMAIN STILL. These sleek, streamlined members of the weasel family are full of energy. They are constantly on the move. They look for food in water or on land. And they often seem to be playing in rivers, lakes, or oceans. They push sticks and leaves along the surface, chase one another underwater, and dive for pebbles. A river otter may tease a beaver by giving its flat tail a playful tug.

There are many kinds of otters. They live nearly everywhere in the world—in all kinds of terrain, except for polar regions and deserts.

The sea otter, which is found along the rocky

▽ *No sleds are needed as river otters glide along a snow-covered hill. Otters bound for a short distance, and then they slide. A muddy slope or a snowbank near the water often becomes a playground for an otter family.*

420

Pausing as it hunts for fish, a ▷ river otter in Yellowstone National Park pokes its head out of an icy lake. When the water is frozen, otters may use holes in the ice so that they can come to the surface to breathe.

▽ Falling snow does not prevent a female river otter and her three large pups from looking for food on the shore of a pond. Otters usually make their dens at the edge of a lake or a stream. Often one tunnel leads directly into the water.

River otter: 28 in (71 cm) long; tail, 14 in (36 cm)

Pacific coasts of North America and Asia, lives mainly in the ocean. In some areas, it may come ashore to sleep. The river otter spends most of the time in streams, rivers, lakes, and marshes. But, as it looks for food or for a mate, it may travel long distances between bodies of water. River otters are found in Europe, Asia, Africa, and North and South America.

With their webbed feet, short legs, and long bodies, otters are well adapted, or suited, to life in the water. They swim easily, flexing their bodies up and down and paddling with their hind feet. Sometimes they also use their front feet. The animals move their powerful tapered tails as they swim.

An otter's flexible body makes it an underwater

421

Sea otter: 43 in (109 cm) long; tail, 12 in (30 cm)

Anchored by strands of kelp, a kind of large seaweed, a sea otter in California (above) rests easily on its back. The kelp keeps the animal from drifting while it naps. Otters often sleep this way in groups (below) along the rocky coasts of the northern Pacific Ocean. When awake, the otters dive for shellfish and fish.

△ *Underwater, a sea otter holds a sea urchin. The otter will make a meal of the sea urchin despite its spines.*

acrobat. Floating lazily in the water, an otter will suddenly twist and dive down. It glides under the surface. Otters turn somersaults in the water. They swim on their backs or on their sides. Sometimes an otter's sleek, dark head pops out of the water as the animal takes a look around.

Like all mammals, otters breathe air, so they must come to the surface often. When they swim underwater, their nostrils and ears close tightly. This helps keep the water out.

Otters can move quickly on land, but they do not look as graceful as they do in water. When they run, they bound along with their backs arched. Whenever possible, they slide in mud or snow.

OTTER

LENGTH OF HEAD AND BODY: 16-47 in (41-119 cm); tail, 10-28 in (25-71 cm)

WEIGHT: 6-82 lb (3-37 kg)

HABITAT AND RANGE: bodies of water in North America, South America, Africa, Europe, and Asia. Sea otter: coastal areas of western North America and parts of Asia

FOOD: fish, shellfish, small mammals, water birds, eggs, frogs, earthworms, and plants

LIFE SPAN: as long as 22 years in captivity, depending on species

REPRODUCTION: 1 to 6 young after a pregnancy of 2 to 11 months, depending on species

ORDER: carnivores

Some otters live in groups, but they usually search for food on their own. Most otters hunt during the day. In areas where there are people, otters may hunt at night. During the day, river otters may sun themselves on rocks.

River otters eat almost anything they can find. They feed on fish, shellfish, and frogs in shallow water and on birds, eggs, small mammals, and plants on land. Otters eat small fish in the water. Large ones are carried ashore.

Sea otters eat many different sea animals—sea urchins, crabs, clams, mussels, squid, octopuses, fish, and abalone. When a sea otter finds a clam, it has a special way to get at the food inside. The otter brings up a rock from underwater. Then it floats on its back with the rock on its chest. Again and again, the otter bangs the clam against the rock until the shellfish breaks open.

After a sea otter has finished eating a mussel or an abalone, it rolls over in the water to wash the bits of shell and food off its fur. It also grooms its fur by nibbling at its coat with its teeth and by pressing its fur with its paws. Grooming helps the fur stay waterproof. As long as the fur is clean, a layer of short underfur traps air and insulates the otter against the cold.

For almost 200 years, people hunted sea otters for their soft, thick fur until few of the animals were left. Today they are protected by law.

Another otter sought by hunters is the giant otter of South America. The longest member of the otter family, the giant otter grows as long as

△ *After a successful fishing expedition, an African clawless otter drags its catch ashore.*

◁ *Once on land, the otter bites off the fish's head with its sharp front teeth. It often uses its long, sensitive fingers to probe under rocks and in mud for such prey as crabs and crayfish.*

▽ *African clawless otter rolls on a riverbank and removes the water from its coat. The animal may nibble at its fur and use its fingers to groom itself.*

Otter

6 feet (183 cm) and weighs about 60 pounds (27 kg). Giant otters live in family groups.

Most otters come ashore to give birth. Females bear two to six young in underground dens sometimes lined with dry grasses and leaves. The newborn are blind, and they are carefully guarded by their mother. A mother teaches her offspring to survive in the water. When young river otters are about two months old, their mother pushes them in. She watches over them until they learn to swim.

Only the sea otter gives birth in the water, usually to a single pup. The mother lies on her back and holds her pup on her chest. She nurses it, plays with it, and teaches it to dive for its food.

In Venezuela, giant otters (right) poke their heads high above the surface of a river. These sleek animals weigh about 60 pounds (27 kg) and grow as long as 6 feet (183 cm). They live in family groups. At left, a young giant otter pauses on land. Its webbed feet help the animal move easily through the water.

▽ *South American otter glides through the water searching for fish. When it dives for prey, its nostrils and ears close tightly to keep water out.*

South American otter: 26 in (66 cm) long; tail, 20 in (51 cm)

Giant otter: 41 in (104 cm) long; tail, 22 in (56 cm)

P

Paca

Common paca: 28 in (71 cm) long; tail, about 1 in (3 cm)

White dots mark the coat of a common paca in Venezuela. The pattern helps hide the animal as it searches for food along wooded riverbanks.

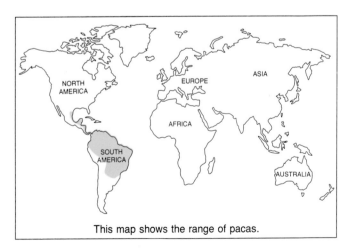

This map shows the range of pacas.

PACA

LENGTH OF HEAD AND BODY: 24-32 in (61-81 cm); tail, about 1 in (3 cm)

WEIGHT: 14-22 lb (6-10 kg)

HABITAT AND RANGE: wooded areas near water from Mexico into South America

FOOD: stalks, leaves, roots, crops, and fallen fruit

LIFE SPAN: about 16 years in captivity

REPRODUCTION: usually 1 young after a pregnancy of about 3 months

ORDER: rodents

LIKE DOTTED LINES, rows of white patches of hair mark the dark coat of the paca. This stocky rodent measures more than 2 feet (61 cm) long. It is one of the largest rodents in the world. It lives in forests from Mexico into Brazil. There its spotted coat helps it hide among forest shadows.

Pacas live alone. During the day, they sleep in burrows. In the evening, they come out to look for leaves, stalks, roots, and fallen fruit.

Pacas frequently make their burrows in riverbanks. With their claws and sharp teeth, they dig burrows that may reach 5 feet (152 cm) in depth. Pacas usually have several openings into their burrows. That way, they have more than one escape route from an enemy, such as a jaguar or an ocelot. The animals swim well. Often a paca will leap into the water to avoid danger. At other times, a paca will snap at an enemy with its teeth. It gives a low, coughing growl if it is angry or alarmed.

Like its close relative the agouti, the paca is especially fond of avocados and mangoes. Read about the agouti on page 47. Near towns and farms, pacas sometimes eat sugarcane, yams, and other crops. People in Mexico and Central and South America sometimes kill pacas for meat and because the animals damage their crops.

Another kind of paca, the mountain paca, lives high in the Andes. Smaller than the common paca, this animal has thicker hair marked with spots.

Both common pacas and mountain pacas usually bear only one young at a time. A female may give birth twice a year.

Panda

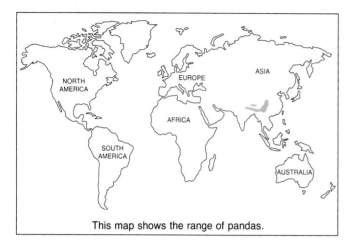

This map shows the range of pandas.

Giant panda: 5 ft (152 cm) long; tail, 5 in (13 cm)

TURNING SOMERSAULTS and standing on their heads, giant pandas are crowd pleasers. Visitors flock to zoos to see the black-and-white animals—among the best-loved in the world.

Though the giant panda's bearlike appearance is familiar to most people, its behavior in the wild is still largely a mystery. Much of what scientists know about pandas has been learned from those in captivity. Fewer than forty pandas live in zoos.

Pandas are difficult to study in their natural surroundings. They live only in remote mountain areas of the central People's Republic of China. There the climate stays cool and wet all year. Snow remains on the ground from fall until late spring. At other times, heavy rains soak the dense bamboo forests where the pandas make their homes. The animals may live as high as 13,000 feet (3,962 m). In the summer, the animals stay on the higher slopes. In the winter, they move down to the valleys.

Though pandas sometimes eat birds and small rodents, their main food is bamboo. A panda consumes huge amounts of the plant. Munching for hours at a time, it may eat as much as 20 pounds (9 kg) of bamboo every day.

With its paw, a panda clutches the tender shoots and leaves of bamboo plants. The animal's long wristbones serve almost as thumbs and help it grasp even thin pieces of food.

When eating, a panda sits down and stretches out its hind legs. Sometimes it rolls backward until its legs stick up in the air. The animal mashes food with its powerful jaws and strong back teeth. As the

Bending a bamboo stalk to its mouth, a giant panda in a zoo in Europe firmly clutches its favorite food. A long wristbone serves almost as a thumb. It allows the panda to grip its food. The panda may eat as much as 20 pounds (9 kg) of bamboo a day. The tough, stringy plant does no damage going down. A thick lining in the panda's throat protects it from splinters. Giant pandas occasionally feed on birds and small rodents.

427

panda swallows, the thick lining of its throat protects it from bamboo splinters.

Occasionally, there are shortages of bamboo. The plants flower rarely, but when they do, the stalks soon die. For several years, the new seedlings are too small for pandas to eat. Because pandas eat little else, some of the animals may starve when food is scarce. No one knows how many pandas remain in the wild today.

Pandas usually live alone. They meet at mating time in the spring. About five months later, a female may give birth to one or two cubs in a cave or other sheltered spot. With its coat of thin, white fur, a panda cub looks like a tiny kitten. By the time it is a month old, it has an adult panda's black-and-white markings. Born with its eyes closed, the offspring cannot see until it is two months old. It crawls at three months of age.

Until the cub is about four months old, it drinks only its mother's milk. Then the young panda begins to nibble bamboo leaves and tender shoots. Although it weighs only about 5 ounces (142 g) at birth, the cub gains weight rapidly. When fully grown, a female weighs about 250 pounds (113 kg). A male may grow to 300 pounds (136 kg). Both measure about 5 feet (152 cm) long.

Many years before the giant panda became known to scientists outside China, another panda had been discovered. Until the mid-1800s, the red panda was thought to be the only kind of panda in the world. Named for the fiery color of its coat, the red panda is the size of a large house cat. It weighs about 12 pounds (5 kg) and measures about 2 feet (61 cm) long. Its bushy, ringed tail adds another 18 inches (46 cm).

The red panda lives in the same kind of habitat as its larger relative. Its range, though, is much wider—from Nepal to northern Burma and into the central part of China.

The red panda spends much of its life in trees. When asleep, it curls up on a branch and wraps its tail around its body. In warm weather, the animal

Big and barrel-shaped, a giant panda lumbers slowly along the ground. The animal swims well, however, and climbs trees with ease. Though a popular sight in zoos, the panda remains a mystery in the wild.

Red panda: 26 in (66 cm) long; tail, 18 in (46 cm)

Striped face of a red panda peers from behind a tree. Some scientists group red pandas and giant pandas together in their own family. Other experts place the smaller red panda in the raccoon family.

stretches out on a limb, legs dangling on either side.

The red panda has a wristbone much like that of the giant panda. Like its larger relative, the red panda eats bamboo. It feeds on fruit, acorns, roots, and perhaps on small animals as well. It grasps the food and carries pieces to its mouth.

Red pandas are shy animals, and they usually live alone. The adults meet only in the mating season. About four months later, the female usually bears one to three young. Their eyes open within three weeks.

◁ *Hsing-Hsing—a gift from the People's Republic of China to the United States in 1972—feeds among stalks of bamboo. He and his female companion, Ling-Ling, live at the National Zoological Park, in Washington, D. C.*

Some scientists put the red panda in the raccoon family and the giant panda in the bear family. Other scientists group the two kinds of pandas together in their own family.

PANDA

LENGTH OF HEAD AND BODY: 5-6 ft (152-183 cm); tail, 5 in (13 cm). Red panda: 20-26 in (51-66 cm); tail, 11-20 in (28-51 cm)

WEIGHT: 165-300 lb (75-136 kg). Red panda: 7-13 lb (3-6 kg)

HABITAT AND RANGE: mountain forests in the central People's Republic of China. Red panda: mountain forests from Nepal to northern Burma and into central China

FOOD: bamboo, grasses, roots, acorns, fruit, and small animals

LIFE SPAN: 13 years in captivity

REPRODUCTION: 1 to 3 young after a pregnancy of 4 to 5½ months

ORDER: carnivores

431

Pangolin

(say PANG-guh-lin*)*

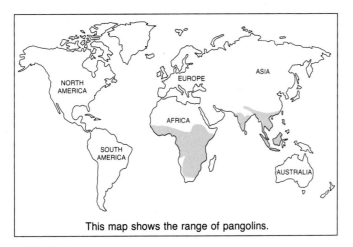

This map shows the range of pangolins.

PANGOLIN

LENGTH OF HEAD AND BODY: **12-31 in (30-79 cm); tail, 12-31 in (30-79 cm)**

WEIGHT: **10-60 lb (5-27 kg)**

HABITAT AND RANGE: **grasslands and forests of Africa and southern and southeastern Asia**

FOOD: **mainly ants and termites**

LIFE SPAN: **as long as 14 years in captivity**

REPRODUCTION: **usually 1 young after a pregnancy of unknown length**

ORDER: **pholidotes**

THE PANGOLIN LOOKS like a pinecone with legs. Tough, overlapping scales cover most of the animal's stocky body. As it moves through the darkness, it walks on all fours, dragging its tail. From time to time, it sniffs the air for food or danger.

Because of their scales, pangolins often are called scaly anteaters. But the animals are not related to other anteaters. Scientists place pangolins in an order, or group, of their own.

Pangolins are born with scales that are small and soft at first. They become harder and larger as the pangolin grows. An adult animal can measure 5 feet (152 cm) long, including its tail.

Pangolins live on grasslands and in forests in Africa and in Asia. There are seven kinds. The larger ones stay mostly on the ground and live in burrows. Smaller pangolins make their homes in trees. Tree pangolins use their strong tails to grip branches as they move about. Sometimes they even hang upside down, holding on only with their tails. Pangolins usually live alone.

Chinese pangolin: 21 in (53 cm) long; tail, 14 in (36 cm)

Cape pangolin: 19 in (48 cm) long; tail, 13 in (33 cm); diameter rolled up, 10 in (25 cm)

△ *Chinese pangolin (top) searches for food as it climbs a tree limb. Tough scales cover most of the animal's body. Some pangolins live in trees. Enemies like lions or tigers cannot reach them there. Other pangolins stay on the ground. They curl up into hard, scaly balls when enemies come near. Two young lions poke at a curled-up Cape pangolin (above) in Africa. Usually only a lion, a tiger, or a person can unroll the armored ball.*

Pangolins search for food mostly at night. When one finds a nest of ants or termites, it claws it open. Then it uses its long tongue to gather the insects from inside the nest. The pangolin has no teeth, so the insects are swallowed whole.

A pangolin usually sleeps all day, curled up in a ball. The hard scales on its back protect its soft, hairy belly. A mother shields her soft-scaled young by curling herself around her offspring. A female pangolin usually has one young a year.

Pangolins also curl up when threatened. Only a strong or a skillful hunter—a lion, a tiger, or a person—can unroll it. Even when a pangolin is not curled up, it has other defenses: Larger kinds use their muscular tails to swat attackers. Other pangolins spray enemies with strong-smelling urine.

Indian pangolin feeds at a termite mound. To get at the food, the animal tears holes in the nest with its thick front claws. Then it pushes its snout inside and extends its long, wet tongue. As a protection against the bites of termites, the pangolin can close its nostrils, ears, and thick-lidded eyes.

Indian pangolin: 24 in (61 cm) long; tail, 18 in (46 cm)

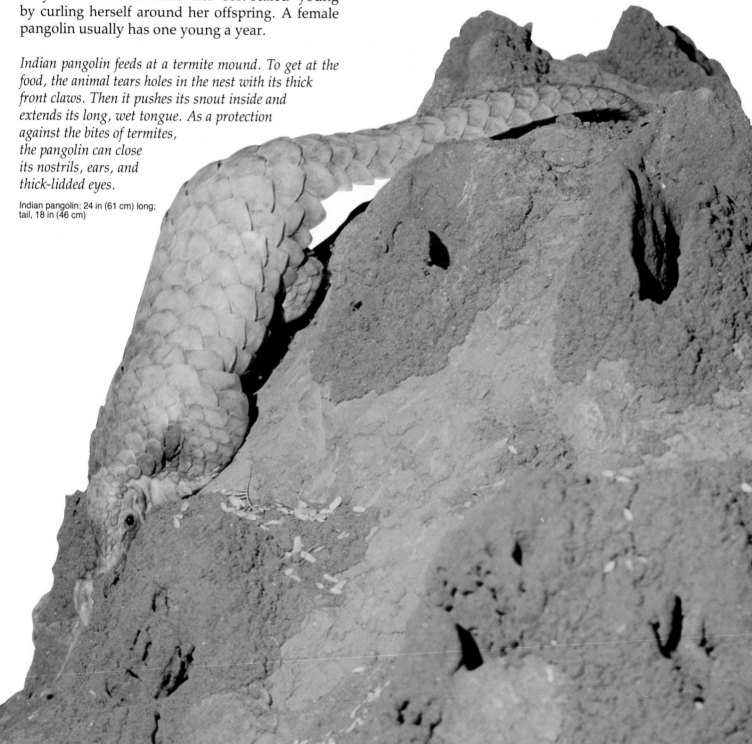

Panther

Panther is another name for the leopard. Read about the leopard on page 329.

Peccary

(say PECK-uh-ree)

Collared peccary: 22 in (56 cm) tall at the shoulder

△ *Sniffing and touching snouts, a young collared peccary and an adult greet each other. Newborn peccaries can travel with the herd only one day after birth. Young animals usually stay very close to adults.*

△ *Collared peccary bites into a prickly poppy flower in Texas. Peccaries use their razorlike teeth for fighting off such enemies as coyotes, wild dogs, and bobcats. Because of their long, sharp teeth, peccaries are sometimes called javelinas, from a Spanish word meaning "spear."*

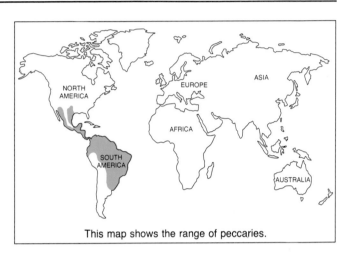

This map shows the range of peccaries.

A HERD OF PIGLIKE ANIMALS steps out of the shade where the group has rested during the midday heat. A dozen peccaries—old and young, male and female—are looking for food in a Texas desert.

Some of the peccaries sniff the ground. They can detect plant bulbs under as much as 6 inches (15 cm) of soil. The peccaries dig out the bulbs with their rubbery snouts. Other peccaries munch cactuses—spines and all. They also eat mushrooms, berries, fruit, acorns, and new shoots of grass.

Peccaries drink from streams or ponds. Unlike some animals that live in deserts, peccaries cannot get all the moisture they need from plants. Their small hoofprints mark the trails that lead to water.

Peccaries are found from the southwestern United States into South America. When resting or feeding, the animals usually keep out of sight among cactuses, small shrubs, or rocks. But the noises they make give them away. The animals snort and grunt softly as they move about. They bark sharply when they fight over food.

Peccaries look and sound much like their distant relatives, wild pigs. They are about the size of small pigs, and they have piglike snouts. Their chunky, short-legged bodies are covered with coarse, salt-and-pepper-colored bristles.

The stiff hairs on the peccary's back hide a scent gland that produces an oily, strong-smelling

Small herd of collared peccaries—young and old, male and female—trots across a grassy field in Texas.

substance. When a peccary rubs its back against a rock, it leaves a scent mark. Other peccaries in the herd recognize the smell because they often rub their faces against one another's backs. Members of a herd keep track of each other by the scent. People can smell the mild skunklike odor from as far as 300 feet (91 m) away.

When an enemy is near, or when a peccary is angry, it raises the bristles along its back. Then it woofs loudly. Alerted, the other peccaries in the herd scoot quickly into the underbrush.

Peccaries are timid and will run away from danger if they can. If attacked, however, they defend themselves fiercely. They bite and slash with razor-sharp teeth. Fighting as a group, the members of a herd can overcome a wild dog, a coyote, or a bobcat. They may even be able to drive away a jaguar.

Once a year, a female peccary may give birth. She bears her young—usually twins—in a sheltered place. The next day she returns to the herd with her offspring. The young peccaries will be protected by all the members of the herd.

Only the smallest peccary—the collared peccary—lives in the United States. It gets its name from a band of light-colored bristles around its neck. Collared peccaries live in deserts in the Southwest and in dry woodlands and dense rain forests in Central and South America.

The white-lipped peccary lives only in tropical rain forests in Central and South America. It may roam in herds of a hundred animals. The largest peccary, the tagua (say TAHG-wah), lives on dry grasslands and in forests in central South America.

PECCARY

HEIGHT: 20-30 in (51-76 cm) at the shoulder

WEIGHT: 30-66 lb (14-30 kg)

HABITAT AND RANGE: deserts, woodlands, and rain forests in parts of North and South America

FOOD: cactuses, grass shoots, bulbs, nuts, berries, flowers, mushrooms, and fruit

LIFE SPAN: about 7 years in the wild

REPRODUCTION: usually 2 young after a pregnancy of about 5 months

ORDER: artiodactyls

435

Phalanger

Greater glider: 15 in (38 cm) long; tail, 20 in (51 cm)

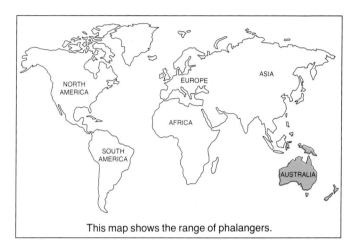

This map shows the range of phalangers.

PHALANGER

LENGTH OF HEAD AND BODY: **3-32 in (8-81 cm); tail, as long as 25 in (64 cm)**

WEIGHT: **1 oz-20 lb (28 g-9 kg)**

HABITAT AND RANGE: **forests of Australia, New Guinea, New Zealand, and neighboring islands**

FOOD: **leaves, fruit, flowers, nectar, sap, and small animals**

LIFE SPAN: **3 to 20 years, depending on species**

REPRODUCTION: **1 to 6 young after a pregnancy of 2 to 5 weeks, depending on species; length of pregnancy unknown for some**

ORDER: **marsupials**

Brush-tailed possum: 18 in (46 cm) long; tail, 12 in (30 cm)

△ *Largest of the gliding phalangers, a greater glider balances on a branch. It soars from tree to tree by extending flaps of skin that connect wrists and ankles.*

Young brush-tailed possum in Australia looks to its ▷ *mother at mealtime. Like other marsupials, phalangers carry their young in pouches while the offspring develop. Brush-tailed possum offspring stay in the pouch for about five months and nurse for several months more.*

436

Feather-tailed glider: 3 in (8 cm) long; tail, 3 in (8 cm)

Wait for me! Tiny feather-tailed glider hurries after its mother as she searches for insects to eat. Smallest of the gliders, this kind of phalanger uses its fringed tail to steer as it glides from tree to tree.

WHEN ENGLISH EXPLORER James Cook first visited Australia in 1770, his men captured a cat-size woolly animal. It had a partially naked tail and a pouch on its belly. It reminded the Englishmen of the American mammal called the opossum. So they gave the same name to the Australian animal. Over the years, the name was shortened to possum to distinguish it from the American opossum.

Scientists have learned that possums and opossums are not closely related. Both animals belong to the group of pouched mammals called marsupials (say mar-soo-pea-ulz). But each animal is a member of a separate family. Australian possums are members of the phalanger family. Their family name comes from a Greek word that means "bone of a finger or toe." The flexible fingers and toes of phalangers help these animals climb. Other members of the phalanger family include the cuscus and the koala. Read about them under their own headings.

Phalangers live in the forests of Australia, New Guinea, New Zealand, and neighboring islands. Some are as small as (Continued on page 440)

Phalanger

Fluffy glider dines on the sap of a eucalyptus tree. ▷
These phalangers are long-distance gliders. They
can sail as far as 300 feet (91 m). Because of the
color of the fur on their bellies, some people call
them yellow-bellied gliders.

▽ Standing on three paws,
a green ring-tailed possum
feeds on a leaf. This animal's
name comes from the greenish
color of its fur and the way its
tail curls up. Ring-tailed
possums usually travel in
pairs. In the trees, they build
ball-shaped nests of
branches and leaves.

Fluffy glider: 10 in (25 cm) long; tail, 15 in (38 cm)

Green ring-tailed possum: 12 in (30 cm) long; tail, 11 in (28 cm)

▽ *Herbert River ring-tailed possum carries her four-month-old young. Females bear two young at a time.*

◁ *Hanging on by tail and toes, a Herbert River ring-tailed possum reaches for a leafy meal. When not dangling from a branch, this possum of northeastern Australia usually keeps its tail curled up.*

Herbert River ring-tailed possum: 12 in (30 cm) long; tail, 11 in (28 cm)

Leadbeater's possum peers from behind a leaf. ▷ *This animal feeds mainly on insects and nectar.*

Leadbeater's possum: 6 in (15 cm) long; tail, 10 in (25 cm)

field mice. Others are about the size of house cats. Most phalangers are skillful climbers. Their sharp claws dig into tree branches, and their toes spread apart for a firm grip. Most have soft, thick fur and long tails. A phalanger may use its tail like an extra hand when it climbs. The animal's tail can curl

▽ *On the prowl, a striped possum searches for insect larvae. With strong teeth and a long claw on each front foot, a "stripey" can dig out larvae beneath tree bark.*

Striped possum: 10 in (25 cm) long; tail, 12 in (30 cm)

Honey possum: 3 in (8 cm) long; tail, 3 in (8 cm)

△ *Honey possum feeds from a flower. A long snout and a bristle-tipped tongue help this animal gather nectar and pollen from blossoms.*

tightly around a tree limb, supporting it until its paws have a grip on the next branch.

Other phalangers have another way of moving through the trees. They glide in swooping dives. These animals are called gliders. They leap into the air and spread out their arms and legs. Flaps of skin extend from their wrists to their ankles. The animals float gently down to another tree. Some of the larger gliders make glides as long as 300 feet (91 m)—the length of a football field.

Phalangers usually bear one to six offspring after a pregnancy of a few weeks. The tiny, hairless newborn crawl into their mother's pouch. There they continue to grow and develop. Within a few months, the young are covered with fur. Then they are able to go into and out of the pouch. Sometimes the mother carries them on her back.

When they are fully grown, many kinds of phalangers go off on their own. Some phalangers, such as honey possums, travel about in pairs. Others, like the gliders, often live in family groups.

Phalangers rest during the day. For shelter, some phalangers build nests of leaves, twigs, and branches. Others sleep curled up in hollow limbs or tree trunks. Some just stretch out along branches, hidden among leafy shadows.

At night, phalangers wake up and move about the trees. They search for such food as leaves, fruit, flowers, nectar, sap, and pollen. One of the gliding phalangers is called the sugar glider because of its fondness for sweet food. Read about this animal on page 530. Some phalangers also eat insects and other small animals.

Phalangers have few enemies. They include snakes, monitor lizards, and large owls. The scratching noise of a lizard climbing a tree makes a phalanger scream in alarm. Some phalangers, like the brush-tailed possum, will fight fiercely. Gliders will leap to another tree. But others, like the slow-moving ring-tailed possum, stay perfectly still. If the enemy does not leave, the possum tries to creep away. It often gets caught.

Climbing upward, a squirrel glider stretches from ▷ *one bare branch to another. On the way down to another tree, the animal may make a daredevil dive. It can glide more than 100 feet (30 m) at a time.*

Squirrel glider: 8 in (20 cm) long; tail, 9 in (23 cm)

Pig

Pig is another name for the hog. Read about hogs on page 264.

Pika

(say PEA-kuh *or* PIE-kuh)

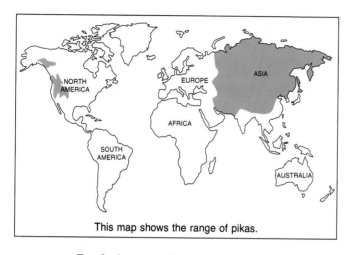

This map shows the range of pikas.

SCURRYING OVER ROCKS on a mountain slope in late summer, the pika carries a mouthful of plants. Soon the small, grayish brown animal reaches the haystack it is building. It climbs to the top and adds the plants it has gathered. In time, the sun-dried heap—the pika's winter food supply—may be several feet high and equally as wide.

Some kinds of pikas live on the rocky slopes of mountains in Asia, southeastern Europe, and western North America. Other kinds make their homes on grassy plains and in some deserts of Asia. From spring until late fall, there are plenty of plants for the animals to eat. The hardworking pikas begin in

Perched on top of its haystack, a pika in Wyoming guards its winter food supply. If another pika tries to steal a leaf or a blade of grass, the pika will promptly chase away the intruder.

Pika: 8 in (20 cm) long

midsummer to pile up grasses, herbs, and twigs. When cold weather sets in and plants become scarce, the haystacks provide food. Pikas stay active in winter. Covered with thick, soft fur, they travel about in tunnels they dig under the snow.

Many pikas live in colonies, or families. Those that live on plains and in deserts make their homes in underground burrows. On mountain slopes, pikas seek shelter among rocks. The animals find high places to use as lookouts. There they watch for danger. If they spot an enemy—a hawk, an eagle, or a weasel—the pikas call out a warning and scamper into rock crevices to hide.

Smaller than their relatives, hares and rabbits, pikas measure about 8 inches (20 cm) long. Read about hares on page 250 and rabbits on page 472.

Although hares and rabbits are usually silent, most pikas make sounds. Pikas that live in some areas have calls that may sound like sharp barks. Pikas in other areas make sounds that resemble the bleats of lambs.

A pika litter usually includes two to six young. When eight days old, the offspring can crawl and make peeping noises. Less than two months after birth, the young pikas reach full size. By then, their mother may have a new litter to care for.

△ *On guard duty, a pika in Montana barks an alert. Its rocky lookout provides a good view of the slope below. After sounding a warning, the pika will dash for cover.*

PIKA

LENGTH OF HEAD AND BODY: 6-10 in (15-25 cm)

WEIGHT: 6-14 oz (170-397 g)

HABITAT AND RANGE: rocky slopes of mountains, grassy plains, and some deserts in western North America, southeastern Europe, and parts of Asia

FOOD: grasses, herbs, and twigs

LIFE SPAN: about 3 years in captivity

REPRODUCTION: 2 to 6 young after a pregnancy of about 1 month, depending on species

ORDER: lagomorphs

▽ *Large leaf makes a midmorning snack for a collared pika in Alaska. Pikas feed on many kinds of plants. In the spring and early summer, they spend much of their time eating. Then pikas begin to pile up food for winter.*

▽ *Rubbing its cheek against a rock, a pika leaves a substance produced by a gland in its face. The pika marks rocks with this scent or with urine to claim an area as its own. Scent marks may also help pikas find mates.*

Collared pika: 8 in (20 cm) long

Platypus

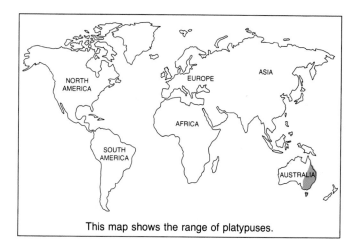

This map shows the range of platypuses.

PLATYPUS

LENGTH OF HEAD AND BODY: about 15 in (38 cm); tail, 5 in (13 cm)

WEIGHT: about 3 lb (1 kg)

HABITAT AND RANGE: near lakes and streams in eastern Australia

FOOD: worms, young shellfish, insects, and insect larvae

LIFE SPAN: 17 years in captivity

REPRODUCTION: usually 1 or 2 young hatched from eggs after an incubation period of about 10 days

ORDER: monotremes

WHEN SCIENTISTS IN BRITAIN first saw a platypus almost 200 years ago, they doubted that it was a real animal. This strange creature does exist in Australia, however. Its forefeet are webbed like those of a duck. Its muzzle looks like a duck's bill. Its tail resembles that of a beaver. And its fur looks like an otter's fur. The animal measures less than 2 feet (61 cm) long from its muzzle to the tip of its tail.

The platypus is a monotreme (say MON-uh-treem). Besides the echidna, it is the only mammal that bears its young by laying eggs. Read about the echidna on page 188.

On land, the platypus shuffles along. In the water, it moves much more gracefully. It dives for food at dawn and at dusk. As the platypus submerges, folds of skin close over its ears and eyes. It also can close off its nostrils.

The platypus can stay beneath the surface for a minute or two at a time. Using its sensitive, skin-covered muzzle, the platypus probes the bottoms of lakes and streams for food. The animal scoops up worms, young shellfish, insects, and insect larvae, along with some mud and gravel. It stores this mixture in its cheek pouches.

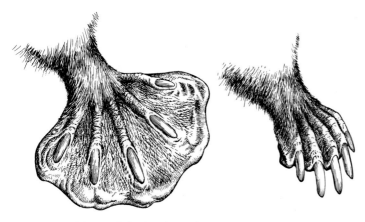

△ *Webbing of the forefoot of a platypus extends beyond its nails and forms a fan-shaped paddle (above, left). The paddle helps push the animal through the water. When the platypus moves on land, the webbing folds back (above, right). Then the platypus can walk or run more easily. It also can use its thick nails for digging.*

▽ *Sleek and streamlined when submerged, an adult platypus swims along a gravelly bottom. It paddles with its front feet and steers with its tail and hind feet. Folds of skin cover its eyes and ears when it is underwater. Its nostrils also can close. The platypus gets food by sifting through mud and gravel with its sensitive muzzle.*

When its pouches are full, the platypus rises to the surface to chew its food. The adult platypus has grinding pads instead of teeth. The mud and gravel in the animal's mouth help crush the food between these pads.

The male platypus has sharp spurs on the heels of its hind feet. To defend itself, it can jab enemies with these spurs and discharge a strong poison.

The platypus lives in a burrow that it digs at the edge of a lake or a stream, just above the water's surface. Using its thick nails and muzzle, it loosens dirt and then packs it to form the walls of its burrow.

A female platypus digs a nesting burrow where she lays her eggs. This winding tunnel usually is about 25 feet (8 m) long. At the end of the tunnel, the female lines a small chamber with grass and leaves. She seals herself into the chamber with a plug of dirt and lays one or two sticky, leathery eggs. She keeps the eggs warm between her body and her tail.

The eggs hatch in about ten days. The hairless newborn platypuses are only about the size of jelly beans. The young suck milk that flows from pores on their mother's belly. After several months, they begin to swim and to dive on their own.

Pocket gopher

Botta's pocket gopher: 6 in (15 cm) long; tail, 3 in (8 cm)

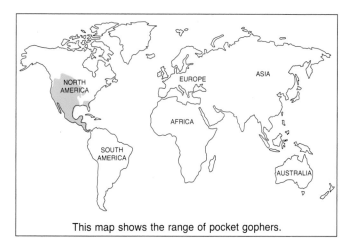

This map shows the range of pocket gophers.

△ *Botta's pocket gopher pops out of its burrow in Arizona for a quick look around.*
▽ *Plains pocket gopher in Minnesota carries food in its bulging cheek pouches. A pocket gopher gets its name from the fur-lined pouches on the sides of its face.*

Plains pocket gopher: 8 in (20 cm) long; tail, 4 in (10 cm)

POCKET GOPHER

LENGTH OF HEAD AND BODY: 4-14 in (10-36 cm); tail, 2-5 in (5-13 cm)

WEIGHT: 3-32 oz (85-907 g)

HABITAT AND RANGE: deserts, open forests, grasslands, valleys, and mountain slopes in parts of North America

FOOD: mainly roots and bulbs

LIFE SPAN: about 2 years in the wild

REPRODUCTION: 1 to 8 young after a pregnancy of 3 or 4 weeks

ORDER: rodents

HURRYING THROUGH ITS BURROW, the pocket gopher patrols its underground home. It looks for food and for intruders—weasels, snakes, and other pocket gophers. About thirty species, or kinds, of pocket gophers live in parts of North America. Most of the time, these stocky rodents, sometimes simply called gophers, live alone in their burrows. If two gophers happen to meet, they may fight to the death.

A pocket gopher, though usually less than 1 foot (30 cm) long, can move an amazing amount of dirt for an animal its size. One gopher's burrow may extend for hundreds of feet. Loosening dirt and stones with its large teeth and claws, the animal digs separate chambers for sleeping, for storing food, and for waste.

Some burrows lie just below the surface of the ground. There the gopher finds roots and bulbs to eat. When it nibbles on the root of a plant, it may pull the stalk underground, too. People have watched entire plants disappear, yanked down by a gopher.

The gopher usually does not eat all its food at once. It cuts up plants with its large front teeth. Then it uses its paws to stuff the food into fur-lined pouches on the sides of its face. The gopher carries the food to its underground storeroom. There it empties its pouches. Like pockets, these pouches can be turned inside out for cleaning.

The only time gophers live together is when they have young. A female usually gives birth once or twice a year. A litter may include as many as eight offspring. The young stay with their mother for only about two months.

Stripes and spots of brown, white, and yellow color the fur of a marbled polecat, a relative of the weasel.

Polecat

(*say* POLE-cat)

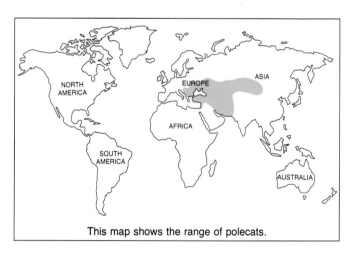

This map shows the range of polecats.

POLECAT

LENGTH OF HEAD AND BODY: 11-14 in (28-36 cm); tail, 5-7 in (13-18 cm)

WEIGHT: about 3 lb (1 kg)

HABITAT AND RANGE: dry, grassy plains and brushy areas in parts of Europe and Asia

FOOD: small mammals, birds, reptiles, and frogs

LIFE SPAN: 8 years in captivity

REPRODUCTION: 4 to 8 young after a pregnancy of 2 months

ORDER: carnivores

ON WINDSWEPT PLAINS in parts of Europe and Asia, the marbled polecat seeks its prey. This relative of the weasel hunts in the morning and evening. It usually feeds on mice, young hares, ground squirrels, birds, frogs, and lizards.

Like some of its relatives—weasels, ferrets, and grisons—the long, slender marbled polecat can slither through narrow openings into underground burrows after prey. The polecat may kill more food than it can eat. It hides what is left over in a burrow and returns to eat it later.

Marbled polecats live and hunt alone, except during the mating season. In an underground nest, a female polecat bears four to eight young in the spring. The young polecats stay with their mother until the summer.

When attacked by enemies such as dogs or foxes, the polecat fluffs its marbled fur and curves its tail over its back. It bares its teeth and growls. Then the polecat may turn around and squirt a smelly liquid, produced by glands under its tail, at its enemy. The polecat's vivid markings may serve as a warning signal for other animals to stay away.

447

Porcupine

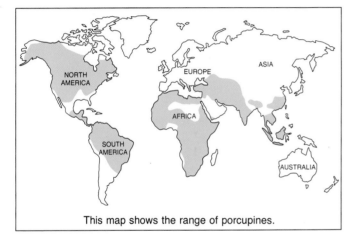

This map shows the range of porcupines.

Sharp point of a quill of a North ▷
*American porcupine (far right) can
easily pierce an attacker's body.
Overlapping scales, or barbs (right),
cover the quill's tip. Like barbs on
a fishhook, they catch in skin and
are difficult to pull out.*

◁ *North American porcupine pauses
while nibbling twigs with its orange
front teeth. Like other rodents, the
porcupine wears down its front teeth by
gnawing on stems and other hard foods.*

▽ *Gripping a branch with its strong, curved claws, a
North American porcupine rests in a birch tree in Alaska.
Pads on its hind feet help it hold on. These animals spend
much of their time in trees.*

WHEN ALARMED, a porcupine may stamp its feet,
click its teeth, and growl or hiss. But a prickly coat of
needle-sharp quills is the animal's best defense. The
pointed quills are stiff, thick spines banded with
black, brown, pale yellow, or white. The quills usually cover the porcupine's back, sides, and tail. They
are mixed in with the animal's softer hairs.

Usually a porcupine's quills lie flat against its
body. But if an enemy such as a bobcat, a hyena, or a
fisher comes too close, a porcupine raises and
spreads its quills. Contrary to a popular belief, the
animal cannot shoot them at an attacker. But the

◁ *Turning its back, a North American porcupine warns
an attacker to stay away. A halo of guard hairs surrounds
its bristling quills. When the porcupine brushes against
an enemy, the quills may lodge in the enemy's flesh.*

North American porcupine: 29 in (74 cm) long; tail, 8 in (20 cm)

Ground dwellers, a pair of crested porcupines looks for shelter in a termite mound in Kenya.

quills are loosely attached. They may come off when a porcupine slaps its tail or brushes against an enemy. New quills grow in to replace lost ones.

The word *porcupine* means "quill pig" in Latin. Porcupines, however, are not pigs. They are large rodents. They live in deserts, forests, and grasslands. There are two groups of porcupines. One group lives in North and South America. These animals differ in looks and habits from porcupines found in Europe, Africa, and Asia. They all shuffle along the ground. But porcupines in North and South America also climb trees. Some can even use their tails to grasp branches.

The only porcupine found in the United States and Canada is the North American porcupine. In winter, the animal feeds on bark and evergreen needles. In spring and summer, it eats leaves, buds, stems, and fruit. In campgrounds, porcupines may gnaw on ax handles or canoe paddles.

A female porcupine bears one to four young after a pregnancy of two to seven months, depending on the species. The newborn can see at birth. Hair and soft quills cover most of their bodies. Within a few days, their quills harden. When the offspring are about two months old, they leave their mother and go off on their own.

Prehensile-tailed porcupine: 18 in (46 cm) long; tail, 16 in (41 cm)

PORCUPINE

LENGTH OF HEAD AND BODY: 12-34 in (30-86 cm); tail, 2-18 in (5-46 cm)

WEIGHT: 2-60 lb (1-27 kg)

HABITAT AND RANGE: forests, deserts, and grasslands in parts of North and South America, Europe, Africa, and Asia

FOOD: bark, leaves, buds, stems, fruit, and sometimes crops

LIFE SPAN: as long as 20 years in captivity

REPRODUCTION: 1 to 4 young after a pregnancy of 2 to 7 months, depending on species

ORDER: rodents

◁ *Prehensile-tailed porcupine skillfully climbs along a tree branch in Bolivia. Its grasping tail coils around another branch.*

▽ *Two bristly bodyguards protect a young crested porcupine in Tanzania. By raising their quills—which may be 1 foot (30 cm) long—they may frighten away an attacker.*

451

Porpoise (say POR-pus)

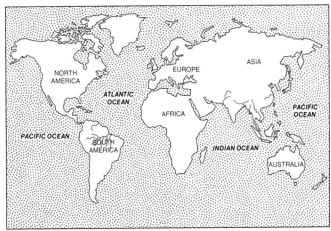

This map shows the range of porpoises and dolphins.

PORPOISE AND DOLPHIN

LENGTH OF HEAD AND BODY: 5-31 ft (152 cm-9 m)

WEIGHT: 75 lb-9 t (34-8,165 kg)

HABITAT AND RANGE: all oceans and some freshwater rivers

FOOD: fish, squid, shrimps, birds, and mammals

LIFE SPAN: 8 to 50 years in the wild, depending on species

REPRODUCTION: 1 young after a pregnancy of 10 to 16 months, depending on species

ORDER: cetaceans

SHOOTING OUT OF THE WATER like a rocket, a porpoise moves through the air in a graceful arc and splashes back into the sea. Another porpoise leaps, and then another. Soon a whole school, or group, of porpoises is taking part in a playful water ballet.

Porpoises live in all the oceans of the world and in several freshwater rivers. Though they may look a little like fish, porpoises are actually mammals. They are small, toothed whales. You can find out about larger whales on page 568. Some porpoises are called dolphins (say DOLL-fins). People often use both words to refer to the same animal.

There are more than fifty species, or kinds, of porpoises. They vary greatly in size and shape. The harbor porpoise weighs 75 pounds (34 kg) and measures 5 feet (152 cm) long. The largest porpoise, the huge orca (say OR-kuh), may weigh as much as

◁ *Porpoise with a friendly grin, a bottlenose dolphin glides through waters off Hawaii. These porpoises frequently perform in oceanariums. People often use the words* dolphin *and* porpoise *to refer to the same animal.*

Bottlenose dolphin: 9 ft (274 cm) long

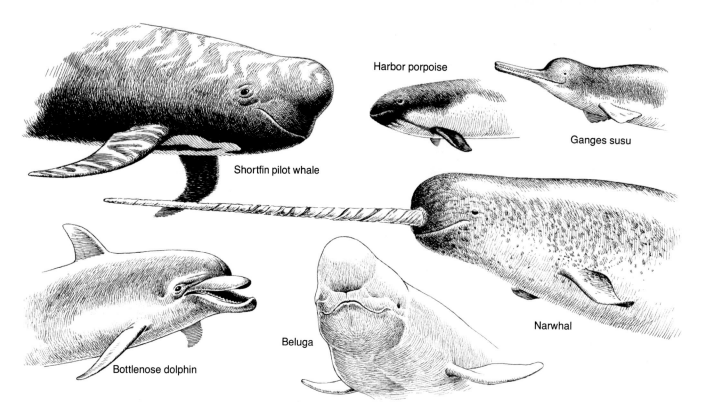

Shortfin pilot whale

Harbor porpoise

Ganges susu

Narwhal

Bottlenose dolphin

Beluga

△ *More than fifty kinds of porpoises live around the world. They swim in several freshwater rivers and in all the oceans. Clockwise, from lower left: The bottlenose dolphin uses its beaklike snout to catch fish in coastal areas. The shortfin pilot whale feeds mostly on squid. A harbor porpoise has a cone-shaped head. It eats herring and cod. The long beak of a Ganges susu helps it probe muddy river bottoms for food. The male narwhal has a long tusk, but scientists do not know what it is used for. The beluga lives in cold, northern waters. To breathe, it may ram holes in newly formed ice with its back.*

Largest of all porpoises, a mighty orca, or killer whale, ▷ *hurls itself into the air. The crashing sound it makes as it splashes down may scare fish into coves and bays. There the orca can hunt them for food.*

9 tons (8,165 kg). It is sometimes called the killer whale, and an average adult measures 25 feet (8 m) in length.

Some porpoises have round heads. Others have long, beaklike snouts. The male narwhal (say NAR-whahl) has an ivory tusk—a long tooth—growing from the tip of its snout. The tusk may measure 8 feet (244 cm) long. Scientists do not know why the narwhal has such a feature, though some think the tusk may help the animal attract mates.

The porpoise must surface from time to time to

Orca: 25 ft (8 m) long

get a breath. The animal has a single nostril, called a blowhole, on the top of its head. Right below this opening is a valve like a plug. When the porpoise surfaces, it contracts a muscle and opens the valve. The animal exhales a breath and inhales another. As the porpoise dives back under the water, it closes the valve again.

Most porpoises come up to breathe about every four minutes. Even when they are sleeping, the animals move to the surface to get a breath. Some, like the bottlenose dolphin, breathe about every twenty seconds. Other kinds of porpoises can hold their breath for as long as thirty minutes.

Porpoises are well suited to their environment. A thick layer of blubber, or fat, protects them from cold temperatures. Their torpedo-shaped bodies are streamlined for swift, easy movement. Flippers on their sides help them balance and steer. Tail fins,

△ *Attack! Hungry orcas speed toward a seal. They will knock it off the ice and into the cold waters off Antarctica. Orcas often prey on other marine mammals.*

called flukes, move up and down. They push the porpoises forward. These fast swimmers often speed through the water. Some kinds can travel more than 25 miles (40 km) an hour.

Porpoises that live in coastal areas have good eyesight. Some that live in rivers, such as the Ganges susu of India, are blind. They depend mainly on their keen sense of hearing to find their way.

Most porpoises navigate by using echolocation (say ek-oh-low-KAY-shun). They also use echolocation to find such food as fish and other sea animals. As a porpoise swims, it makes a series of clicking sounds that travel through the water. When the sounds hit an object, echoes bounce back to the porpoise. By listening to the echoes, a porpoise can tell the size, shape, and location of the object. And it can tell if the object is moving. Using echolocation, porpoises can stay out of the way of enemies. They also can avoid obstacles such as icebergs or boats. Bats also use echolocation to find food and to avoid obstacles. Read about bats beginning on page 77.

Porpoises also may avoid attackers because of their coloring. Most porpoises have dark backs. From above, they are hard to spot in the murky

◁ Dusky dolphin sails through the air off the coast of Argentina. Its companions speed alongside. Duskies swim in schools of as many as a hundred animals.
▽ Shortfin pilot whale in the Pacific Ocean leaps straight out of the water. These big porpoises get their name from their habit of following a leader, or pilot— usually the largest animal in the group.

Shortfin pilot whale: 22 ft (7 m) long

Dusky dolphin: 8 ft (244 cm) long

depths. Seen from below, their light bellies blend with the brighter water near the surface.

Swimming together in large numbers helps give porpoises additional protection against danger. Schools may range in size from a few animals to perhaps 10,000. While swimming in schools, porpoises communicate with each other. They whistle, squeak, growl, and moan. When in trouble, porpoises sometimes cry out for help.

Porpoises have been known to go to the rescue of a sick or injured porpoise. Often the distressed animal has difficulty swimming to the surface to

Beluga: 15 ft (5 m) long

△ *Belugas, or white whales, cruise the Arctic Ocean near Somerset Island off Canada. In these clear, shallow waters, more than a thousand belugas come together to give birth.*

Porpoise

breathe. The others may support it with their flippers and swim with it to the surface. Porpoises help one another in other ways, too. When a female porpoise gives birth, nearby females may assist her.

About a year after mating, a pregnant female gives birth to one young, usually in early spring. If the newborn does not rise immediately for its first breath of air, it may drown. The mother and the other females gently push it to the surface.

The mother nurses her young for about ten months. Special muscles allow her to squirt her milk into its mouth.

Intelligent and curious, porpoises have been taught to do many tricks. They often perform at oceanariums. There they jump through hoops, play catch, and snatch fish from a person's hand. Scientists are teaching these animals many things. Someday, they hope, porpoises may learn a language of signs and symbols. If so, they will be able to communicate with humans.

Air bubbles rise from a beluga's blowhole as it sounds ▷ off underwater. All porpoises make whistling noises, but belugas also growl, roar, and squeal.

Potto

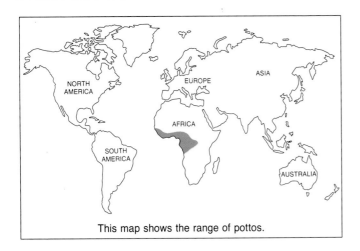

This map shows the range of pottos.

POTTO

LENGTH OF HEAD AND BODY: 12-16 in (30-41 cm); tail, 2-3 in (5-8 cm)

WEIGHT: about 2 lb (1 kg)

HABITAT AND RANGE: forests in central and western Africa

FOOD: fruit, tree gum, insects, and snails

LIFE SPAN: 7 to 9 years in captivity

REPRODUCTION: 1 young after a pregnancy of about 6 months

ORDER: primates

▽ *Long-fingered and stubby-tailed, a potto clings to a branch with a powerful grip. It uses the claws on its hind feet to groom its fur. These small primates climb slowly through the trees of central and western Africa.*

NOISELESSLY, the potto climbs through the forests of central and western Africa at night. The squirrel-size animal usually moves very slowly and carefully. It lets go with one foot only after the other three have firmly gripped a branch. The loris, an Asian relative of the potto, moves in a similar way. Read more about lorises on page 348.

Pottos, like monkeys, apes, and humans, are members of the primate order. Their thumbs resemble those of other primates, and pottos can use their thumbs and fingers to grip strongly.

Clinging to branches, pottos sleep the day away. Their bodies are curled into tight woolly balls. When they wake up at dusk, pottos begin to feed on fruit, tree gum, and insects. Their keen noses and eyes help them find food in the dark.

Under the sensitive skin on the back of the potto's neck are several long spine bones. Scientists are not sure what the bones are for. Some think that they may help the animal defend itself. When alarmed, a potto may tuck its head between its front legs and butt against its enemy. If necessary, it can defend itself by biting with its sharp teeth. But the potto's best defense is remaining unnoticed.

A potto has one young each year. The offspring nurses for about two months. Then it begins to feed on fruit. When it is grown, it will go off on its own.

Prairie dog

(say PRAYER-ee DOG*)*

△ Black-tailed prairie dog remains on the lookout as it feeds. The small rodent eats grasses, seeds, and roots.
◁ Head thrown back, a black-tailed prairie dog calls a signal to other members of its town. Bison and prairie dogs often share an area. Sometimes the huge bison roll in the dirt on prairie dog mounds.

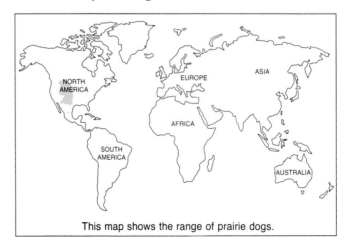

This map shows the range of prairie dogs.

Black-tailed prairie dog: 14 in (36 cm) long; tail, 3 in (8 cm)

ACROSS THE PRAIRIES of North America, people sometimes see mounds of firmly packed dirt. The mounds mark the bustling towns of black-tailed prairie dogs. Among the mounds, a hundred plump, rabbit-size rodents may scurry about. Some search for grasses, roots, and seeds. Others busily shape the entrance mounds to their underground burrows. A few may sun themselves. As they move around, the prairie dogs stay alert to danger.

Enemies of prairie dogs can often spot their prey on the flat, open prairies. But the prairie dogs have an alarm system to protect themselves. When one of the small rodents senses danger, it may sit up and cry a warning that can be heard throughout the town. Hearing the alarm call, other prairie dogs quickly dash for safety and vanish inside their underground homes.

After a few minutes, several prairie dogs will cautiously peer from their burrows. If the danger has passed, they may give an all-clear signal. They come out of their burrows, rear on their hind legs, and call out loudly. Gradually, more of the animals emerge. They scamper about the prairie and carry on their daily activities.

Inside a prairie dog burrow are one or more long tunnels a few feet below the ground. Several chambers, or rooms, lie along each tunnel. One of the chambers may serve as a nursery where prairie dog young nestle in a bed of dry grasses. Another chamber is a listening post. From that room, prairie dogs can hear what is happening above ground.

The entrance mound of the burrow provides a

△ *Young prairie dog peeks over the entrance mound to its burrow after hearing an all-clear signal. Black-tailed prairie dogs also have special calls that warn of danger, help them claim territory, and express contentment.*

lookout station for prairie dogs. It also keeps water from flooding the burrow. Prairie dogs form the mound with loose dirt. They spend a great deal of time keeping it in good shape. They kick dirt onto the mound and then pack the soil with their noses and foreheads. Around some mounds, there is actually a pattern of noseprints.

Some people who live on the Great Plains think prairie dogs are a nuisance. The rodents dig holes in fields and in pastures. So people sometimes kill prairie dogs. Many black-tailed prairie dogs remain on

Rebuilding after a storm, a busy prairie dog shapes an entrance mound. First it kicks loose dirt onto the mound (above, left). Then it rams its nose against the soil (above, center) to pack it down. Covered with dirt, the animal sits in front of its rounded ring of earth (right) and looks around the prairie dog town.

the Great Plains, however. The resourceful rodents continue to dig their burrows.

Black-tailed prairie dogs live in towns that include many small groups of animals. Each group is usually made up of an adult male, several females, and young born that year. Every group has its own territory, or area. If a strange prairie dog approaches, the residents try to chase the intruder away. Members of the group share food, play together, and groom one another. When they meet, they kiss or nuzzle each other. In these ways, the animals can identify members of their group.

A prairie dog town may be home to hundreds of other animals. Snakes and burrowing owls may take shelter in the burrows. Coyotes, eagles, and other enemies hunt the prairie dogs in the towns.

A few black-footed ferrets may also live there.

Enemy of black-tailed prairie dogs, a coyote prowls ▷ through the grass. If the coyote comes too near, the prairie dogs will duck into their burrows.

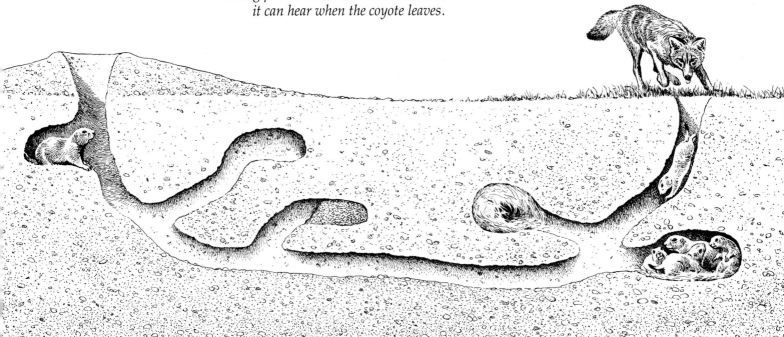

Underground homes, prairie dog burrows provide places to sleep, to raise young, and to find shelter. Too quick for a coyote, a prairie dog quickly disappears into a hole. In a nursery chamber (below, right), a mother huddles with her pups. Nearby is a sleeping chamber with a nest. Along the tunnel (from right to left), the prairie dogs also have built a toilet chamber and a higher, dry chamber that rarely floods. Another prairie dog stations itself at a listening post close to the mound entrance. There it can hear when the coyote leaves.

PRAIRIE DOG

LENGTH OF HEAD AND BODY: 12-15 in (30-38 cm); tail, 3-4 in (8-10 cm)

WEIGHT: 2-4 lb (1-2 kg)

HABITAT AND RANGE: prairies and mountain plains in parts of the western United States and Mexico

FOOD: grasses, seeds, roots, and leafy plants

LIFE SPAN: about 8 years in captivity

REPRODUCTION: about 5 young after a pregnancy of 1 month

ORDER: rodents

△ *Relaxing in a field of wild flowers in South Dakota, a prairie dog family basks in the June sun. The mother stands watch while one of her pups nurses.*

◁ *Prairie dog carries a mouthful of grass to line its nursery chamber. This nest inside the burrow will provide a warm, safe place to sleep and to raise young.*

The number of ferrets decreased during the last century as people tried to get rid of prairie dogs—the ferrets' prey. With fewer prairie dogs around, the black-footed ferret suffered from lack of food. Today very few of the animals may still roam the prairie. Read about ferrets on page 202.

Less numerous than black-tailed prairie dogs are their white-tailed relatives. White-tailed prairie dogs make their homes on high mountain plains farther to the west than the towns of black-tailed prairie dogs. The white-tailed animals live in smaller groups. Instead of building large towns, the animals may dig widely scattered burrows with only a few other prairie dogs nearby. White-tailed prairie dogs make fewer calls than their relatives do.

During the winter months, both white- and black-tailed prairie dogs usually stay underground. They live mostly on the fat stored inside their

△ *Do I know you? Prairie dog pups nuzzle each other. Small groups of black-tailed prairie dogs live together and share a territory. By touching and sniffing, the animals identify members of their own group.*

△ *In a friendly romp, two young prairie dogs take a tumble. Pups may play together when they meet. Adults often stop to kiss and to groom one another.*

bodies. White-tailed prairie dogs hibernate (say HYE-bur-nate), or sleep, for as long as six months. Their body temperatures drop, and their heart rates slow down. Black-tailed prairie dogs may wake up on warmer days and look for food outside.

In late winter or early spring, female prairie dogs may give birth to tiny offspring. An average litter includes about five young, called pups. Until the pups are about six weeks old, their mother nurses them. Then they leave the safety of the burrow and begin to eat grasses and leafy plants. Soon they are frisking playfully with other young prairie dogs.

Pronghorn

(say PRONG-horn)

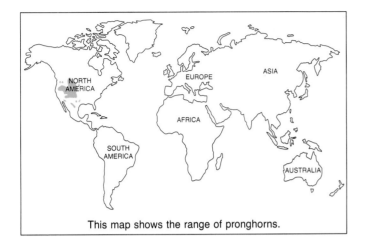

This map shows the range of pronghorns.

◁ *Dark-eyed pronghorn doe alertly watches a prairie in northern Mexico. Large eyes and ears alert her to distant movements and sounds. The creamy-white bands on her throat, mouth, and cheeks contrast with the brownish color of her coat.*

▽ *Pair of pronghorns trots across a rolling grassland in the Black Hills region of South Dakota. The male, or buck, wears the pronged horns that give these animals their name. The female, or doe, has only bony knobs on her forehead.*

FASTEST MAMMAL IN NORTH AMERICA, the pronghorn can sprint more than 50 miles (80 km) an hour. At half that speed, this graceful, deer-size animal can run for several miles. Pronghorns feed and travel in small bands or in large herds. Throughout many parts of western North America, groups of pronghorns leap and gallop across the plains.

Males, or bucks, have two horns, pronged and curved at the tips. Females, or does, have smaller horns without prongs. Some does have no horns at all. Each fall, the outer covers of the horns drop off, leaving only skin-covered, bony cores. By summer, the coverings have grown back—even larger.

On older animals, the pronged horns may measure more than 1 foot (30 cm) long. Because of the horns, some people believe that pronghorns are related to antelopes. Scientists disagree, however, on whether the animals are closely related.

Pronghorns eat mostly sagebrush and other shrubs that grow in the dry areas where they live. Where water is scarce, pronghorns can survive for a long time on moisture they get from plants.

Pronghorns swallow their food without chewing it thoroughly. After eating, they bring up a wad

Pronghorn

Newborn pronghorn lies hidden in prairie grass in ▷ South Dakota. Its mother is probably grazing a short distance away. She stays close to her offspring that way, but she does not reveal its hiding place.

▽ Sniffing the air for the scent of a coyote or a bobcat, a pronghorn buck in Montana snorts an alarm. The hairs of his white rump patch stand on end and serve as a warning signal to pronghorns farther away.

of partly digested food, called a cud. They chew the cud, swallow it, and digest it completely.

Pronghorns are always alert to such enemies as coyotes or bobcats. Pronghorns have keen eyesight. Because their eyes are on the sides of their heads, they can keep watch in most directions.

When the pronghorn is alarmed, hairs on its neck and rump stand on end. The flash of white hair on its rump acts as a signal to other pronghorns. As the pronghorns speed away, glands near their tails produce a strong-smelling liquid. The scent of this liquid also helps warn of danger.

The pronghorn has a brownish coat with white markings. The hairs in its coat are coarse and about 2 inches (5 cm) long. During cold weather, these hairs lie flat. They hold body heat next to the pronghorn's skin, helping to keep the animal warm. In hot weather, the pronghorn flexes certain muscles, and the hairs in its coat stand up. Air flows among the hairs and cools the animal.

A buck has a dark mane on his neck and black

PRONGHORN

HEIGHT: **about 3 ft (91 cm) at the shoulder**

WEIGHT: **90-150 lb (41-68 kg)**

HABITAT AND RANGE: **grasslands and shrubby areas of western North America**

FOOD: **shrubs and grasses**

LIFE SPAN: **about 7 years in captivity**

REPRODUCTION: **usually 2 young after a pregnancy of about 8 months**

ORDER: **artiodactyls**

masklike markings on his head. In the mating season in September, he marks his territory with a liquid from a gland in his face. A buck then selects several does. Herding the females together—and chasing other bucks away—keeps him busy. Rival bucks snort and fight. Sometimes they injure each other with their horns.

In the spring, the does go off by themselves to give birth, usually to twins. A mother hides her offspring among tall grass. After a few weeks, the young pronghorns can run with their mothers. They will be fully grown after a year and a half.

▽ *Surrounded by sagebrush, a favorite food, pronghorns move from one grazing area to another. Pronghorns feed on the low shrubs and grasses that grow in the dry, open areas where they live.*

Puma

Puma is another name for the mountain lion. Read about it and other cats on page 126.

Quokka

IN THE COOL EVENING, quokkas leave the shallow, grassy resting places where they have slept most of the day. They move along well-traveled pathways through swamps and thickets in southwestern Australia and on two nearby islands. The animals stop often to nibble tender shoots. If frightened, they quickly hop away. With their short hind legs, quokkas spring through the underbrush.

Quokkas are small members of the kangaroo family—only about 3 feet (91 cm) long from head to tail. They are marsupials (say mar-soo-pea-ulz), or pouched mammals, like many other mammals in Australia. Read about kangaroos on page 310.

Short, coarse, grayish brown hair covers the quokka's stocky body. Its thin tail is nearly hairless and not very long. Its ears are small and rounded.

Each quokka usually stays within a home range that may overlap those of other quokkas. Sometimes a group of more than a hundred quokkas lives in a large area made up of many overlapping home ranges. Members of one group rarely enter another group's area.

Quokkas can go for a long time without drinking water. They get most of the moisture they need from the plants they eat. Even so, many of the animals do not survive during very hot, dry summers.

Mating occurs during the winter. One tiny, underdeveloped offspring is born about a month later. It crawls into its mother's pouch, where it grows larger and stronger. About five months after birth, the young quokka begins to leave the pouch. At six months of age, the animal goes off on its own. But it remains in the same area as its mother.

Quokkas are also known as wallabies. Read about other wallabies on page 554.

◁ *Paws drawn toward her chest, a bright-eyed female quokka rests standing up. Thick, coarse hair covers her body and hides the opening to her pouch.*

Hungry quokkas nibble seeds scattered over sandy soil. ▷ *When feeding, quokkas move slowly about on all fours. But if frightened, they quickly hop away.*

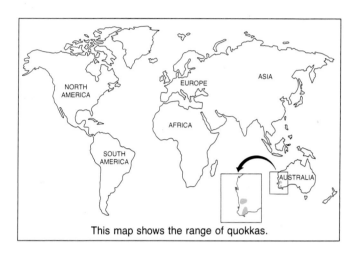

This map shows the range of quokkas.

QUOKKA

LENGTH OF HEAD AND BODY: 19-23 in (48-58 cm); tail, 10-14 in (25-36 cm)

WEIGHT: about 7 lb (3 kg)

HABITAT AND RANGE: brushy areas of southwestern Australia and nearby islands

FOOD: grasses and other plants

LIFE SPAN: about 5 years in captivity

REPRODUCTION: 1 offspring after a pregnancy of about 1 month

ORDER: marsupials

R

Rabbit

EVEN IF YOU LIVE IN A CITY, you may have spotted the white flash of a rabbit's tail as the small mammal raced for cover. If you are sharp-eyed, you may have spied a rabbit as it crouched—absolutely still.

Rabbits are among the most familiar mammals in the world. There are nearly 25 species, or kinds, of rabbits. They live on all continents of the world except Antarctica. They inhabit swamps, marshes, deserts, woodlands, grasslands, prairies, and volcanic slopes. Some rabbits are found near people in cities and suburbs.

European rabbits live together in a complex burrow system called a warren. Using their forepaws and powerful hind feet, the rabbits dig out several openings and a network of tunnels and chambers. In North America, most rabbits spend the day resting in shallow dens called forms. But when danger threatens, these rabbits may seek shelter in the abandoned burrows of other animals.

The rabbit looks very much like its close relative the hare. But rabbits are usually smaller than

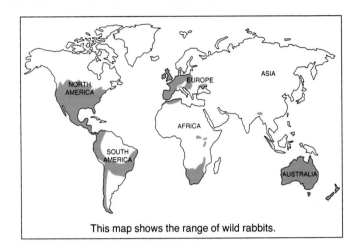

This map shows the range of wild rabbits.

Ears alert, a Nuttall's cottontail remains still as it ▷ *watches for enemies in a snowy field in South Dakota. Winter or summer, Nuttall's cottontails are most active in the morning and the evening.*
▽ *Desert cottontail in southern Arizona samples a bright yellow flower on a brittlebush plant. Found in many kinds of habitats in southwestern North America, this kind of rabbit feeds mainly on grasses and herbs.*

Desert cottontail: 13 in (33 cm) long; tail, 2 in (5 cm)

Nuttall's cottontail: 12 in (30 cm) long; tail, 2 in (5 cm)

hares—only about 16 inches (41 cm) long and 3 pounds (1 kg) in weight. Rabbits also have shorter ears and shorter hind legs than hares do. Newborn rabbits start life helpless, without hair on their bodies, and with their eyes closed. Young hares, on the other hand, can leap about almost as soon as they are born. They have fur, and their eyes are open. Read about hares on page 250.

Like hares, rabbits eat many kinds of plants. They nibble on grasses and herbs. When grass is scarce in winter, the animals gnaw on twigs or bark. Rabbits that live near people sometimes eat vegetables and other crops.

In fields, rabbits usually can find plenty of grasses and herbs to eat. But open areas can be dangerous places for rabbits. The list of the rabbit's enemies is long. Foxes, dogs, bobcats, coyotes, lynxes,

weasels, raccoons, hawks, and eagles all prey on rabbits. And people hunt them for sport, for food, and for fur. Rabbits usually defend themselves by running away. Sometimes they may freeze, or sit very still, trying not to be seen. European rabbits may race into the safety of their warren. Some rabbits may swim to escape danger. Others, like the desert cottontail and the brush rabbit, may even climb trees. In spite of all these defenses, life is very difficult for rabbits in the wild. Most live for only about a year. If rabbits did not have so many young every year, the small mammals might have become extinct long ago.

Different kinds of rabbits mate at different times throughout the year. During courtship, male rabbits, called bucks, become rivals for the females, called does. European bucks may fight by boxing with their forepaws. They also strike at each other with their powerful hind feet. At mating time, males and females seem to be very affectionate. They may lick each other's ears and heads.

About a month after mating, a doe digs a shallow den and lines it with fur and grasses. There she bears two to seven tiny offspring, often called kittens. Each kitten weighs less than 2 ounces (57 g). It looks more like a little hairless guinea pig than like a rabbit. The doe covers the nest with grass to protect the young from heat and cold. The grass also helps

RABBIT

LENGTH OF HEAD AND BODY: **11-23 in (28-58 cm); tail, 1-5 in (3-13 cm)**

WEIGHT: **1-16 lb (454 g-7 kg)**

HABITAT AND RANGE: **all kinds of habitats in parts of every continent, except Antarctica; domestic rabbits live in many parts of the world**

FOOD: **grasses, herbs, twigs, and bark**

LIFE SPAN: **about 1 year in the wild**

REPRODUCTION: **2 to 7 young after a pregnancy of about 1 month**

ORDER: **lagomorphs**

European rabbit: 16 in (41 cm) long; tail, 2 in (5 cm)

Newborn European rabbits (far left) lie naked and helpless in a nest lined with fur and grasses in England. By four days of age (left, center) the young rabbits—called kittens—are much larger and have fine, thin fur. They still spend most of the time curled up asleep. After about fourteen days, the bright-eyed kittens have become more active. Like the kitten at left, they begin to groom their own fur. By the time the kittens are about one month old, they may begin to explore outside. Below, an adult peers from the entrance to its warren. If it spots danger, the rabbit will quickly disappear down the hole.

hide the kittens during the day, when their mother is away from the nest. Each night, the mother rabbit returns to the nest and nurses her young. The kittens gain weight quickly, doubling their size in the first ten days.

Within a week, the young rabbits have grown coats of short, soft fur. By the time they are a month old, they have begun to play outside the nest and can find their own food. Offspring are ready to breed in less than a year. By then, their mother may have had other litters. A female rabbit may bear as many as thirty young a year, in about four litters. People often make pets of rabbits. There are more than fifty breeds of domestic, or tame, rabbits and hundreds of varieties. They range in size from the Netherland Dwarf rabbit, at about 2 pounds (1 kg), to the Flemish Giant rabbit, at about 16 pounds (7 kg). Domestic rabbits may have short, plush fur like that of the Rex rabbit or long, woolly fur like that of the Angora (say ang-GORE-uh) rabbit. All domestic rabbits are descendants of the wild European rabbit. They belong to a single species.

Domestic rabbit: 11 in (28 cm) long; tail, 1 in (3 cm)

Marsh rabbit: 16 in (41 cm) long; tail, 1 in (3 cm)

△ *Marsh rabbit nibbles on a leaf in Everglades National Park, in Florida. These rabbits live near the water's edge. There they walk through the soft mud, instead of hopping as other rabbits do. They swim well.*

◁ *Small domestic rabbit sits in a meadow in Germany. Domestic, or tame, rabbits come in many sizes and colors. This one weighs about 2 pounds (1 kg). The largest kind of domestic rabbit may weigh eight times as much.*

Raccoon

(say rack-KOON)

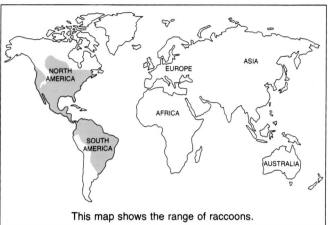

This map shows the range of raccoons.

RACCOON

LENGTH OF HEAD AND BODY: **18-26 in (46-66 cm); tail, 9-12 in (23-30 cm)**

WEIGHT: **7-26 lb (3-12 kg)**

HABITAT AND RANGE: **forests, marshes, prairies, and urban areas in many parts of North and South America and on a few tropical islands**

FOOD: **small animals, including frogs, shellfish, crayfish, worms, mice, and insects, as well as fruit, nuts, and vegetables**

LIFE SPAN: **about 6 years in the wild**

REPRODUCTION: **1 to 7 young after a pregnancy of 2 months**

ORDER: **carnivores**

MANY ANIMALS AVOID PEOPLE and the places they live—but not the raccoon. Its masked face and bushy, ringed tail are familiar to city dwellers as well as to campers in the wilderness. The raccoon is often found in forests near water. It also makes itself at home in marshes, prairies, cities, and suburbs in many areas of North and South America.

Raccoons thrive because they are adaptable—they are able to change their behavior to suit their situation. They will eat almost anything. With their nimble front paws—and a great deal of curiosity—raccoons can find whatever food is available, according to the season. Raccoons have a keen sense of touch. They are always poking their fingers into crevices, searching for small animals, such as mice and insects, to eat. They can even open garbage cans

Masked faces peer from a hollow tree as a female North American raccoon and her six-month-old cub poke their heads out from their den. Raccoons make dens in many sheltered spots—from hollow logs to the attics of houses.

Young North American raccoons explore a hole in a ▷
*sycamore tree in Ohio. A litter of cubs—as many as
seven—stays with its mother for several months.*

and live well on the contents. Near farms, they help themselves to fruit, vegetables, and grain.

Raccoons find much of their food in the water. They feel under rocks and in the mud for crayfish and frogs. A captive raccoon may carry food to its water dish. It seems to lose the food in the water and then to find it again by feeling in the dish with its paws. For a long time, people thought raccoons were washing their food when they did this. The scientific name of the raccoon even means "the washer." Now scientists think that captive raccoons are really acting the way they would in the wild by "finding" food in the water.

Raccoons sleep anywhere that seems safe. A hollow tree or the attic of a building may be used as a den. Often a raccoon has several resting spots within its home range, the area it roams and knows well. The raccoon stays in this area as it searches for food at night. Usually raccoons try to avoid one another. In places where there are many raccoons and plenty to eat, however, several animals may feed together. But they keep their distance from one another.

In early spring, raccoons find mates. About two months later, a female gives birth to one to seven young, called cubs, in a den—often a hole in a tree. Newborn raccoons do not have dark masks on their faces or rings on their tails. These markings appear within a few days.

As the weeks pass, the cubs become more active and the nest becomes crowded. After about two months, the cubs are ready to start exploring their surroundings. Their mother moves them from a den high in a tree to one on the ground. That way, if the cubs take a tumble, they won't be hurt. Soon the

With long, nimble fingers, a young North American ▷
*raccoon pulls persimmons to its mouth. The animal can
skillfully use its paws to investigate an object.*

◁ *Dwarfed by their tree-trunk home, two raccoon cubs
in the state of Washington watch their mother return
from hunting for food. Like their relatives—coatis and
ringtails—raccoons have ringed tails.*

North American raccoon: 24 in (61 cm) long; tail, 10 in (25 cm)

cubs go everywhere with their mother. The family stays in touch by using many sounds—purrs, churrs, twitters, and growls.

During the summer, the cubs become more independent. They often spend several days away from their mother and each other. Raccoons that live in the north eat as much food as they can find during the autumn. Both cubs and adults gain a great deal of weight. By winter, the young rejoin their mother for a few months. The family spends most of that time asleep together in a den. Only on mild winter nights do the animals emerge to look for food. The rest of the time they live on fat stored in their bodies.

A long, freezing winter is hard on young raccoons. Many starve because they do not have as much fat stored up as their parents do. In the spring, a raccoon may weigh only half as much as it did in the fall. By summer, the year-old cubs have found their own home ranges. In warmer areas, raccoon cubs leave their mothers at any time of the year. When food is plentiful and available all year long, cubs reach their full size earlier. Then they can go off on their own sooner.

There are seven species, or kinds, of raccoons. Five of them live on tropical islands. The raccoon that is most familiar roams throughout North America. It can weigh as much as 26 pounds (12 kg). The crab-eating raccoon lives in Central and South America, usually near water. With its sturdy teeth, it can eat hard food such as crabs and other shellfish.

Raccoon in Texas (above) searches for food in a stream. Another animal in Florida (right) licks its paws dry after eating. Raccoons eat almost anything they can find— fish, berries, nuts, mice, eggs, and even garbage.

◁ *Lazily draped among branches, a raccoon (top, left) takes a nap. Adaptable animals, raccoons live comfortably in forests, marshes, prairies, cities, and suburbs. In Arizona, a mother raccoon (left) carries her month-old cub to her den among the rocks.*

Raccoon dog

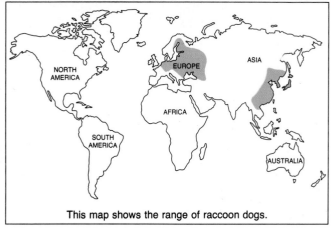

This map shows the range of raccoon dogs.

RACCOON DOG

LENGTH OF HEAD AND BODY: 19-31 in (48-79 cm); tail, 6-10 in (15-25 cm)

WEIGHT: 9-22 lb (4-10 kg)

HABITAT AND RANGE: forests and river valleys in parts of Asia and Europe

FOOD: small mammals, fish, birds, frogs, insects, fruit, nuts, and grain

LIFE SPAN: 10 years in captivity

REPRODUCTION: 5 to 7 young after a pregnancy of about 2 months

ORDER: carnivores

Behind raccoonlike masks, two raccoon dogs prowl a grassy area in Japan. These small wild dogs usually search at night for mice, fish, insects, and other food.

DARK SPOTS AROUND THE EYES give this small wild dog a raccoonlike mask—and its name. Slim in summer, the raccoon dog fattens up for the icy winter by eating large amounts of food. As cold weather sets in, the animal's brownish fur gets longer and thicker. Bundled up in this heavy coat, the raccoon dog looks fat and roly-poly.

Raccoon dogs once were found only in the Far East. There they were prized for their winter fur. During the last century, some were released in western Russia. As the animals bred in the wild, some were trapped for their fur. Wild raccoon dogs have since spread into other parts of Europe.

Nose to the ground, a raccoon dog hunts for food at night. It eats mice, fish, birds, frogs, and insects. It also gobbles berries, nuts, and grain.

In some places, raccoon dogs may sleep through the coldest weather. A pair may curl up together in a moss-lined burrow. Young, usually five to seven pups, are born in early summer.

Rat

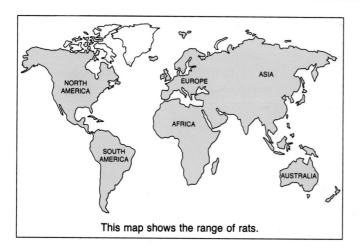

This map shows the range of rats.

RAT

LENGTH OF HEAD AND BODY: 6-23 in (15-58 cm); tail, 3-18 in (8-46 cm)

WEIGHT: 5 oz-15 lb (142 g-7 kg)

HABITAT AND RANGE: almost all kinds of habitats worldwide

FOOD: seeds, nuts, grains, vegetables, fruit, eggs, fish, insects, birds, and other meat

LIFE SPAN: usually less than 1 year in the wild

REPRODUCTION: 1 to 12 young after a pregnancy of 3 weeks to 2½ months, depending on species

ORDER: rodents

FEELING THE WAY with its long whiskers, the rat scurries about, searching for something to eat. With its pointed nose, rounded ears, and long tail, it is a familiar sight. Like its close relative the mouse, the rat is one of the world's most numerous mammals. Read about the mouse on page 398.

There are hundreds of species, or kinds, of rats. Most rats are bigger than mice. Some kinds of rats may measure as long as 23 inches (58 cm), not counting their tails. They may weigh as much as 15 pounds (7 kg). Most rats have scaly tails. But those of the bushy-tailed wood rats of western North America are hairy.

Most rats have short brown, black, or gray hair with light-colored bellies. Their coats may be soft or coarse. But a few rats have special coats. Some kinds of spiny rats in Central and South America have stiff, sharp-pointed hair. The bushy-tailed cloud rat of the Philippines has long hair on its body and tail.

Like mice, rats live in all kinds of habitats—forests, mountains, deserts, and grasslands. Rats live in cities and in the country. Some rats, like the Australian water rat, can swim like a beaver. Others,

▽ *Black-footed tree rat perches on a stump in Australia and flicks its white-tasseled tail. Active at night, these rats usually spend the day in hollow trees.*

Brown rat: 9 in (23 cm) long; tail, 7 in (18 cm)

△ *On a farm in England, a brown rat nibbles stalks of wheat stored after a harvest. One brown rat can eat about 25 pounds (11 kg) of grain in a year.*

Black-footed tree rat: 12 in (30 cm) long; tail, 13 in (33 cm)

like the black rat, can climb trees and jump with ease. Rats make their nests between rocks and in houses, trees, bushes, and burrows.

Some rats build their own homes. Wood rats of North America sometimes pile sticks, twigs, or parts of cactuses against a rock or a bush. Inside, the rats dig out several passageways and chambers. The stick-nest rat of Australia piles sticks together. It sometimes places stones on top to keep the nest from being blown away by strong winds. Inside the nest, the rat is safe from enemies. A stick-nest rat's home may measure 4 feet (122 cm) across and 3 feet (91 cm) high.

The two most common rats—the black rat and

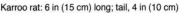

Karroo rat: 6 in (15 cm) long; tail, 4 in (10 cm)

◁ *Emerging from its burrow, a karroo rat watches for enemies. A shy animal, the karroo rat rarely travels far from its home. At a hint of danger, it dashes underground. The karroo rat, also known as the whistling rat, lives in open sandy and grassy regions in southern Africa.*

▽ *Using a wet log as a dining table, an Australian water rat sniffs at its meal—a fish. These rats live near rivers, lakes, and marshes. Their broad, partly webbed feet help them swim after their prey.*

Australian water rat: 11 in (28 cm) long; tail, 11 in (28 cm)

Desert wood rat: 6 in (15 cm) long; tail, 5 in (13 cm)

◁ *Leaving the safety of its rocky home, a desert wood rat in California goes in search of material for its nest. Other wood rats live in nests in trees or on the ground. Wood rats collect all kinds of objects—especially shiny ones—for their nests. Sometimes an animal will drop one object to pick up another. Because of these habits, the wood rat has other names: pack rat and trade rat. People have found glass, cans, silverware, and even mousetraps in the nests of wood rats.*

the brown rat—often live near people. The black rat is also known as the roof rat. It nests in upper stories of buildings as well as in trees. The brown rat, also known as the Norway rat, lives in lower parts of buildings. It is found under floors, in cellars, and in sewers. Although it often lives in filthy surroundings, the brown rat is actually a clean animal. It grooms its fur carefully.

When it lives away from cities, the brown rat digs burrows that include separate living and eating areas. Tunnels connect these chambers and lead to the surface. In their burrows, brown rats live together in family groups called colonies. Usually there are about fifty rats in a colony. But as many as two hundred may stay together. The members of a brown rat colony recognize one another by scent. If a strange rat tries to enter the burrow, the intruder is usually chased away.

Most brown rats are born in nests made of soft, shredded material. Young rats generally nurse for about three weeks. Then they begin to leave the nest to search for food. At about three months of age, they are ready to have their own young.

Some female brown rats may produce seven

On a grassland in Australia, two paler field rats ▷ *huddle as they look for food. Unlike black rats and brown rats, which often live in towns and cities, paler field rats usually live away from people.*

litters in one year, with as many as 12 young in each litter. Very few rats live longer than a year. But a female brown rat can have many offspring. Therefore a colony can grow and remain large.

At night, a brown rat usually leaves its burrow to search for food. Brown rats will eat almost anything, including meat. They sometimes kill such animals as chickens, mice, and fish. Or they will eat meat that people have stored.

Other kinds of rats also eat meat. The cotton rat of North and South America sometimes eats eggs and young birds. The Australian water rat feeds on fish, frogs, water birds, and shellfish. The rat can

Paler field rat: 6 in (15 cm) long; tail, 4 in (10 cm)

White-faced spiny rat: 8 in (20 cm) long; tail, 7 in (18 cm)

White-faced spiny rat slips over a log in South America. The tails of spiny rats break off easily. Scientists believe this helps the animal escape if an enemy grabs it by the tail.

crush some mussels in its jaws. It must place others in the sun. The mussel dies in the heat, and the shell opens. Then the rat can eat its meal.

All rats eat grain and other plants. Every year, these rodents eat or spoil much of the grain grown in the world. Some kinds of rats are very harmful. The brown rat and the black rat cause millions of dollars in crop and property damage in the United States and in other parts of the world. The lesser bandicoot rat destroys many crops in fields and storerooms in India and in neighboring countries. The Polynesian rat also is a pest in parts of Southeast Asia and on some islands in the Pacific Ocean.

Rats do further damage by gnawing. Like all rodents, they have front teeth that grow throughout their lives. They must gnaw to keep their teeth from growing too long. Besides chewing plants, rats sometimes gnaw and damage books, furniture, metal pipes, and electrical wires.

Rats carry many diseases, including typhus and rabies. But rats sometimes are helpful as people search for cures for sicknesses. Scientists often use the animals in medical experiments in laboratories.

Most kinds of rats avoid people and do little damage to crops. Karroo (say kuh-ROO) rats live together in burrows that they dig in plains in southern Africa. During the day, they leave their burrows to feed on nearby plants. The shy animals rarely travel far from their homes. When an intruder comes too close, karroo rats quickly return underground.

Wood rats build their nests far from cities—in trees, on the ground, or in rock crevices. These rats are often called pack rats because they carry off shiny objects such as nails, silverware, glass, and cans. Wood rats use their finds to build their nests. A wood rat often may be carrying one object when it sees something it likes better. It will put the first object down and take the second one, as if it were making a trade. Because of that habit, the wood rat is also called the trade rat.

Another rat that collects objects is the giant pouched rat of Africa. It picks up pens, earrings, keys, or other small things it finds. It takes them to a storeroom in its burrow by carrying them in its teeth or in pouches in its cheeks.

Rats of all kinds are the prey of many other animals, including snakes, owls, weasels, cats, dogs, coyotes, hawks, and foxes.

Ratel

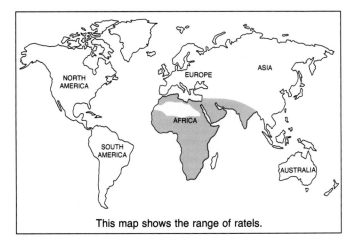

This map shows the range of ratels.

RATEL

LENGTH OF HEAD AND BODY: 24-30 in (61-76 cm); tail, 6-12 in (15-30 cm)

WEIGHT: 15-29 lb (7-13 kg)

HABITAT AND RANGE: many kinds of habitats in parts of Africa and Asia

FOOD: honey, fruit, and large and small animals, including insects

LIFE SPAN: 24 years in captivity

REPRODUCTION: 1 to 4 young after a pregnancy of 6 months

ORDER: carnivores

SNIFFING AT A HOLLOW TREE, a ratel grunts and growls. Soon the black-and-gray relative of the badger finds its goal—a nest of bees. Because it eats honey, the ratel is often called the honey badger.

When hunting for its favorite food, the ratel of Africa may have a helper. A small bird called a honey guide sometimes leads the ratel to a bees' nest. Then the bird waits for the ratel to claw open the nest. After the larger animal has eaten its fill of honey and bee larvae, the honey guide swoops in for the larvae and wax that are left. The ratel's thick skin helps protect it from bee stings.

Besides honey, the ratel feeds on snakes and small mammals as well as on insects and birds. Sometimes a ratel will kill the young of larger animals, such as wildebeests or African buffaloes.

Ratels are found in parts of Asia as well as in Africa. The animals have scent glands that produce a strong-smelling substance. Ratels travel alone, in pairs, or in small groups. They often curl up in rock crevices or in burrows that they have dug with their sharp front claws. There the animals give birth to young. A female ratel probably bears one to four offspring each year.

With its sharp claws, a ratel pokes into a crack in a dead tree. It searches for ants and termites to eat.

Reedbuck

The reedbuck is a kind of antelope. Read about antelopes on page 52.

Rhinoceros

(*say* rye-NOSS-uh-russ)

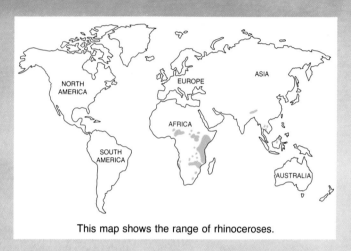

This map shows the range of rhinoceroses.

▽ *Tufted tail raised, a black rhinoceros bull splashes through a shallow lake in Tanzania. Except for females with young, black rhinos live alone—in the mountains and on the dry plains of Africa.*

WITH A SPEAR ON ITS NOSE and thick gray hide covering its body, the rhinoceros looks as tough and protected as an army tank. By the time it is fully grown, the huge plant eater has few enemies except people. Most other animals avoid the powerful adult rhinoceros—one of the largest land mammals in the world.

Rhinoceros means "nose horn." Some kinds of rhinos have one large horn curving upward from their snouts. Other kinds have two. The hard horn, made of a hairlike substance, grows from the rhino's skin. Like hair, the rhino's horn keeps growing—as much as 3 inches (8 cm) a year. The longest known rhino horn measured more than 5 feet (152 cm). If a horn is broken off, a new one will grow.

In Africa, female rhinoceroses use their horns

to defend their young. The rhinos fight off attackers by hooking and butting with their horns.

There are five species, or kinds, of rhinoceroses. Two kinds live in Africa: the black rhino and the white rhino. Despite their names, both of these two-horned animals are gray.

The white rhinoceros may be as tall as 6½ feet (198 cm) at the shoulder. It weighs about 5,000 pounds (2,268 kg). The animal lives on grassy plains in herds of about a dozen animals. White rhinos carry their square muzzles close to the ground. Their heads nod as they walk and graze. Closing the hard edges of their lips tightly, they nip off the grass.

Black rhinos are smaller than white rhinos. Adults live alone in many parts of Africa—in dry inland areas, along the coast, and in the mountains. Instead of grazing, the animals browse on trees and bushes. Wrapping their pointed upper lips around twigs, leaves, thorns, and occasionally fruit, they pull the food into their mouths.

Black rhino: 5 ft (152 cm) tall at the shoulder

Both white and black rhinos feed at dawn, at dusk, and at night. When the noonday sun beats down, the animals lie in the shade or roll in the dust of dry riverbeds. When water holes are nearby, they cool off by wallowing in mud. The coating of mud or dust on their skin helps keep off insects and protects the large animals from the sun.

The Indian rhino lives in Nepal and northern India. The thick skin of this one-horned animal

---

RHINOCEROS

HEIGHT: 3½-6½ ft (107-198 cm) at the shoulder

WEIGHT: 2,240-5,000 lb (1,016-2,268 kg)

HABITAT AND RANGE: grasslands, shrubby areas, and dense forests of Africa and southern and southeastern Asia

FOOD: shrubs, leafy twigs, and grasses

LIFE SPAN: about 40 years in captivity

REPRODUCTION: 1 young after a pregnancy of 8 to 19 months, depending on species

ORDER: perissodactyls

◁ Cattle egret perched on its armored back, a black rhino looks like a rock in the darkness. Sighing and rumbling, the sleeping giant frequently shifts position in the tall grass. The bird catches insects that the rhino disturbs.

▽ Close to its mother's side, a white rhino calf stays safe from lions and hyenas. A rhino mother defends her young furiously. If an enemy approaches, a female white rhino charges and tries to toss it into the air with her horn.

◁ Heavy heads nod as white rhinos cross a grassy plain in Africa. Among the largest land animals in the world, these calm giants feed in groups of as many as a dozen animals. They clear paths to their grazing grounds by walking through high grass or dense brush.

White rhino: 6 ft (183 cm) tall at the shoulder

looks like a suit of armor. Between stiff sections of hide are folds of thinner, more flexible skin. The folds allow the animal's body to move.

On its way to grazing grounds, the Indian rhino travels tunnel-like paths through grass that grows 25 feet (8 m) high. To feed, it curls its pointed upper lip around tall grass stems. It bends the stems over and bites them off. The Indian rhino has large, sharp teeth. A female defends her young by swinging her head and slashing with the teeth in her lower jaw.

In Indonesia, on the western tip of the island of Java, lives another one-horned rhino. It once ranged over much of southeastern Asia. But today only about fifty animals remain in an isolated preserve on the island.

Another kind of rhino lives on the island of Sumatra and in other parts of southeastern Asia. The Sumatran rhino has two horns. The smallest of all rhinos, it measures about 4 feet (122 cm) high at the shoulder and weighs about 2,240 pounds (1,016 kg). The Sumatran rhino has bristlelike hairs on its body. In captivity, the rhino may grow a shaggy coat.

Other rhinos are nearly hairless, except for tufts at the tips of their ears and at the ends of their tails.

Rhinoceroses look awkward, but they are surprisingly nimble and quick. The Indian rhino may charge at 30 miles (48 km) an hour. Rhinos can jump, twist, and turn quickly. Thick, spongy pads cushion

491

Rhinoceros

△ *One-horned Indian rhino feeds on tall elephant grass. The animal grasps the food with its pointed upper lip. Then it bends the stem over and bites it off.*

In the rainy dawn, Indian rhinos wade in a forest ▷ pond. Mud covers their armored hides. When dry, the mud forms a crust that helps protect the animals from sun and insects. Indian rhinos feed during the coolest parts of the day. To escape the midday heat, they wallow chin-deep in water (lower right).

the animals' feet as they move. Rhinos depend on their keen senses of hearing and smell. Often one rhino finds another by sniffing along its trail.

The huge animals communicate with many sounds. They snort, snarl threateningly, and roar. Fighting rhinos grunt and scream. Males and females court with whistling noises.

Every three or four years, a female rhino, called a cow, bears a single calf. On the calf's nose is a smooth, flat plate where its horn will grow. Two-horned rhino calves have two plates. The playful newborn frisks and runs. A rhino cow comforts her young with soft mewing noises. She fiercely defends it from such enemies as lions, hyenas, and crocodiles. The young rhino goes off on its own shortly before its mother bears another calf.

Indian rhino: 5$\frac{1}{2}$ ft (168 cm) tall at the shoulder

Ringtail

DURING THE GOLD RUSH in the American West, prospectors often kept ringtails in their camps to catch rats and mice. The animal's slender body, long whiskers, and big appetite for small rodents earned it other names: miner's cat and coon cat.

This expert ratcatcher is not a cat at all. The dark rings on its bushy tail mark it as a member of the raccoon family. The animal's most striking feature has led to its common name, ringtail. The tail—measuring as long as 17 inches (43 cm)—may be longer than the ringtail's head and body.

Active at night, the ringtail roams forests, deserts, and canyons in parts of the western United States and Mexico. It feeds on small animals, fruit, and plants. This skilled climber easily darts up and down trees. Its hind feet turn backward, and the animal can grasp the bark with its sharp claws when it goes down a trunk headfirst.

During the day, the ringtail rests in a sheltered spot. It lives alone, except during the mating season. As many as four young are born in the spring.

The cacomistle (say KACK-uh-miss-ul) is a slightly larger relative. It spends most of its time in trees in tropical regions in Mexico and Central America.

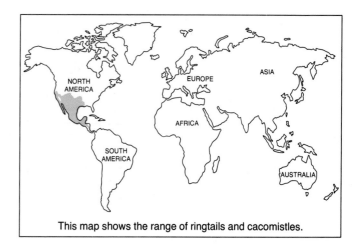

This map shows the range of ringtails and cacomistles.

RINGTAIL AND CACOMISTLE

LENGTH OF HEAD AND BODY: 12-20 in (30-51 cm); tail, 13-20 in (33-51 cm)

WEIGHT: 24 oz-3 lb (680 g-1 kg)

HABITAT AND RANGE: deserts, canyons, and forests from Oregon to Panama

FOOD: rodents, insects, birds, fruit, and plants

LIFE SPAN: 10 years in the wild

REPRODUCTION: 1 to 4 young after a pregnancy of about 2 months

ORDER: carnivores

Bushy, banded tail ▷ dangles as a ringtail rests on a narrow limb in Arizona. An expert climber, the ringtail uses its sharp claws for gripping and its tail for balancing. The tail may be longer than its head and body.

◁ Bright-eyed ringtail pokes its head out of a hollow log. Inside, it has lined a den with grass, leaves, and moss. The opening is just large enough for the ringtail to squeeze through. Ringtails often sleep in crowded spaces in small caves or between rocks.

Ringtail: 12 in (30 cm) long; tail, 17 in (43 cm)

Saki
The saki is a kind of monkey. Read about monkeys on page 376.

Seal and Sea lion

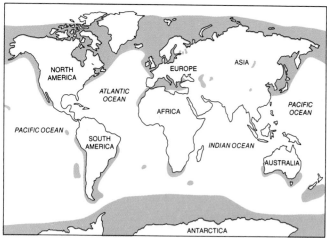

This map shows the range of seals and sea lions.

AMONG THE STAR PERFORMERS at many circuses are trained California sea lions. These sleek, playful seals balance balls on their noses, bark on command, and jump through hoops. In the wild, California sea lions live in the ocean off the rocky coast of western North America and around the Galapagos Islands off South America. The sea lion is one of more than thirty kinds, or species, of seals that inhabit seacoasts throughout much of the world. The largest seal—the southern elephant seal—weighs as much as 8,000 pounds (3,629 kg). An adult ringed seal may weigh as little as 145 pounds (66 kg).

Seals feed on fish, squid, seabirds, and other sea animals called krill that they catch with their pointed teeth. They have keen eyesight and excellent hearing. Not all seals have ears that can be seen, however. Some have only tiny holes in the sides of

Crabeater seals stretch out on an iceberg. By moving their front flippers in a circle and twisting their bodies, these seals can skim the ice at 15 miles (24 km) an hour!

Crabeater seal: 8½ ft (259 cm) long

Seal and Sea lion

Nose to nose, a female harp ▷
seal and her pup sniff each
other. This young white-
coated seal will begin to shed
its baby fur soon after birth.
Gradually, its light coat will
turn dark.

▽ *Young harbor seal swims*
with its mother off the coast of
California. Pups are born on
the shore. They can swim
immediately, and sometimes
they must. The next tide may
cover their birthplace.

Harp seal: 6¹/₂ ft (198 cm) long

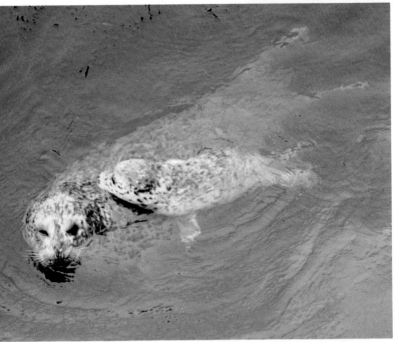

Harbor seal: 6 ft (183 cm) long

their heads. These animals are called earless seals or hair seals. Other seals—called eared seals or sea lions and fur seals—have small ears.

Many kinds of seals live in cold, icy water. A thick layer of fat, called blubber, lines their skins and protects them from low temperatures. Like all marine mammals, seals must swim to the surface to breathe. Weddell seals keep holes open in the ice by gnawing with their sharp teeth.

A few kinds of seals live in warmer waters. Rare monk seals, for example, are found near the Hawaiian Islands and in the Mediterranean Sea.

Seals spend most of the time in the water. Many kinds may not come ashore for weeks at a time. Seals are natural and graceful swimmers. A harbor seal swims by holding its front flippers flat against its body and moving its back flippers from side to side. Sea lions, however, paddle with their front flippers. They use their back flippers only to steer.

When on land, sea lions and fur seals move by lifting themselves up and walking on all four flippers. Hair seals, such as crabeater seals, can only wriggle along on their stomachs. Surprisingly, the crabeater is the fastest seal on land. It can slide along the ice as fast as 15 miles (24 km) an hour.

For most of the year, fur seals swim alone or in small groups. But during the breeding season they come ashore in great numbers. On the islands off the coast of Alaska, older males, called bulls, stake out territories, or areas. There as many as forty females may crowd together. The bulls bark and growl as each fights to control his territory. Younger males without territories gather elsewhere.

During this time, a female fur seal, or cow, bears a single pup. A few days later, the cow mates with a bull. The next year, *(Continued on page 503)*

In shallow waters, a Hawaiian monk seal swims past a ▷
coral reef. Today only about 600 monk seals remain in
Hawaii. Laws now protect these endangered animals.

Hawaiian monk seal: 7¹/₂ ft (229 cm) long

Seal and Sea lion

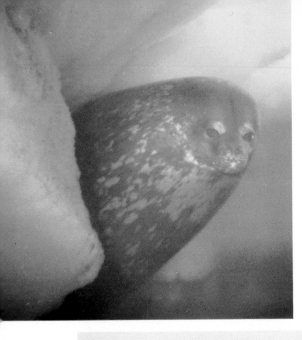

◁ *Snug beneath the ice, a Weddell seal finds shelter in antarctic waters. Closing its nostrils to keep water out, a Weddell seal can dive as deep as 1,900 feet (579 m). It can hold its breath for as long as an hour.*
Weddell seal: 11 ft (3 m) long

Leopard seal clutches a penguin with its teeth. Skillful hunters, leopard ▷ seals prey on seabirds, grabbing them from below as they rest on the water.

▽ *Southern elephant seals crowd together among clumps of grass on an island off South America. Largest of the seals, male elephant seals may grow more than 20 feet (6 m) long and weigh 8,000 pounds (3,629 kg). They get their name not because of their size but because of their trunklike snouts.*

△ *Young northern elephant seals play in waters off Mexico. Their snouts may grow 15 inches (38 cm) long.*

SEAL AND SEA LION

LENGTH OF HEAD AND BODY: 4-20 ft (122 cm-6 m)

WEIGHT: 145-8,000 lb (66-3,629 kg)

HABITAT AND RANGE: coastal waters throughout much of the world, especially in the polar regions, and some freshwater lakes in Asia

FOOD: fish, krill, squid, octopuses, shellfish, and seabirds

LIFE SPAN: 17 to 46 years in the wild, depending on species

REPRODUCTION: 1 young after a pregnancy of 7 to 12 months, depending on species

ORDER: pinnipeds

she will return to give birth again. She nurses her young for four months.

Most newborn hair seals nurse for two to six weeks. A hair seal cow does not take care of her pup for very long. The offspring must survive on its own after only a few weeks.

Polar bears, sharks, and large porpoises called orcas all prey on seals, especially the young. When attacked, a seal may try to defend itself by biting. Or it may quickly dive deep or hide in a hard-to-reach place between rocks or under ice.

People kill seals for their fur or for their blubber. For centuries, some kinds of seals, such as monk seals and northern fur seals, were hunted until they were almost extinct. Today laws help protect these animals.

◁ *South African fur seals gather along a beach in Namibia at the beginning of the mating season. The animals communicate by making a wide variety of noises: honks, bleats, growls, and roars.*

▽ *Underwater acrobats, Australian sea lions glide gracefully through the Indian Ocean. These playful marine mammals have flexible, streamlined bodies.*

▽ *Female northern sea lion barks a threat at a male nearly three times her size. The male, called a bull, controls a small area of land on the rocky coast of Alaska. Within this territory lives a harem of as many as thirty females, or cows. The bull will mate with them, and the cows will each bear a single pup the following year.*

Northern sea lion: 10 ft (305 cm) long

Australian sea lion: 6$\frac{1}{2}$ ft (198 cm) long

South African fur seal: 7$\frac{1}{2}$ ft (229 cm) long

503

Serow

(say suh-ROW)

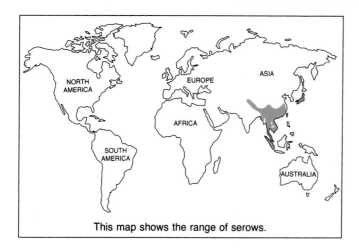

This map shows the range of serows.

SEROW

HEIGHT: 22-39 in (56-99 cm) at the shoulder

WEIGHT: 110-309 lb (50-140 kg)

HABITAT AND RANGE: wooded mountainous areas in parts of Asia

FOOD: grasses, herbs, leaves, shoots, and twigs

LIFE SPAN: as long as 10 years in captivity

REPRODUCTION: usually 1 young after a pregnancy of about 7 months

ORDER: artiodactyls

IN THE RUGGED MOUNTAINS in parts of Asia, the serow picks its way across steep, thickly wooded slopes. Like its close relative the chamois, the serow rarely loses its footing. Short, pointed horns grow from the heads of both males and females, and rough coats cover the animals' goat-size bodies. An adult serow's hair may range from black to red. Some kinds of serows have manes and beards.

Serows may rest much of the day among rocks, in caves, or in dense underbrush. They feed on grasses, herbs, leaves, shoots, and twigs.

Adult serows often live alone in a territory, or area. Each animal marks out its territory by rubbing rocks and branches with a sticky substance produced by glands under its eyes.

Serows mate in the fall. About seven months later, the female gives birth, usually to one young, called a kid. A male may stay with the female and kid for a few months. At about a year old, the kid may go off on its own.

Japanese serow: 22 in (56 cm) tall at the shoulder

◁ Chest-deep in snow, a Japanese serow and her kid find dry leaves to eat. A female serow usually gives birth to one kid. The young stays with its mother for about a year.

Thick woods line a steep ▷ gorge, home of a Japanese serow and her kid. Serows— surefooted relatives of chamois—live in many Asian countries. They clamber easily along well-worn paths on mountain slopes. They often rest among rocks, in caves, or in dense underbrush.

504

Serval The serval is a kind of cat. Read about cats on page 126.

Sheep

HIGH ON A RIDGE, two Rocky Mountain bighorn sheep prepare to battle. Both are males, called rams. They show off their large, curved horns. They growl and kick at each other. Then the rams walk away, turn to face one another, and rise on their hind legs. Suddenly, they lunge, charging at 20 miles (32 km) an hour. Their horns clash, making a loud bang. The sound can be heard a mile away. The rams' stocky, grayish brown bodies actually compress as they come together. After standing still for a moment, the rams repeat the display and clash again.

The battle may continue for hours. But often a fight ends after four or five charges. One ram finally turns away. By giving up, he recognizes the winner as a stronger ram.

Bighorn sheep live in the Rocky Mountains from New Mexico into Canada. Their close relatives,

Rocky Mountain bighorn sheep: 40 in (102 cm) tall at the shoulder

desert bighorns, are found from the southwestern United States into Mexico. Smaller and thinner than Rocky Mountain sheep, desert bighorns roam where few large animals can survive.

Two other kinds of wild sheep inhabit North

From a ridge in Montana, Rocky Mountain bighorn sheep watch for coyotes and mountain lions. Male sheep, called rams, roam together in herds. Rams usually stay with the females only during the mating season.

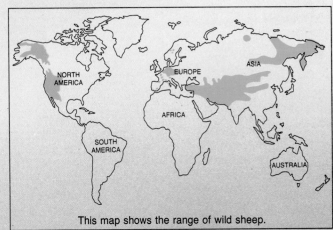

This map shows the range of wild sheep.

America. White Dall's sheep and grayish black Stone's sheep live in the mountains of Alaska and western Canada. Both have slender, widely curving horns that are lighter than those of bighorn sheep. People sometimes call them thinhorns. Other kinds of wild sheep are found in Europe and in Asia.

Wild sheep are relatives of wild goats, and the animals somewhat resemble each other. Like goats, many sheep can move quickly and easily on uneven ground. Their hooves spread out to help them keep their footing. The rough bottoms and hard outer edges usually do not slip as the animals leap from rock to rock. Read about goats on page 232.

All wild rams—and most females, called ewes—have horns. Unlike a deer's antlers, which

△ *In a snowstorm, two young Rocky Mountain bighorn rams fight to determine which is stronger. They rise on their hind*

drop off each winter, a sheep's horns grow longer every year. Most of the growth takes place in spring and summer, when the sheep can find plenty to eat. The deep grooves across a ram's horns show where they stopped growing each fall. By counting the grooves, it is possible to find out a ram's age.

A ram's curving horns can show other details of the animal's life. Broken and splintered tips are the results of fights with other rams. Horns that form a complete circle show that the ram is old. A bighorn ram's horns may weigh 30 pounds (14 kg)—perhaps more than all the bones in his body. The horns of a ewe are smaller than those of a ram.

Like other wild sheep, male and female bighorns and thinhorns live apart for most of the year.

◁ *In Yellowstone National Park, a Rocky Mountain bighorn lamb nibbles tiny plants called lichens from a rock.*

Sheep

Rams roam in herds with other rams. Ewes remain with their lambs and other ewes. In the fall, bighorn rams begin to gather in larger groups. At that time of the year, the number of fights increases. The older rams with larger, heavier horns generally win.

Usually only the strongest rams mate with the ewes when the males and females meet in November and December. After the mating season, the male and female herds separate. When the snow falls, the animals descend to lower mountain slopes.

Older rams lead the male group to the feeding grounds, which may be 25 miles (40 km) away.

Bighorn ram in Montana munches seeds. Though ▷ bighorns usually feed on grass, they also eat other plants.

legs (above, left) and lunge straight at each other (above, center). The rams try to crash head-on. But the horns slip sideways (above, right) and hit the shoulders. In contrast to older rams, the horns of these animals are still small.

Females sometimes travel, but usually for shorter distances. At their feeding grounds, the sheep nibble plants that poke through the snow. Or they may paw through deep, soft drifts to reach the grasses underneath. They may also look for places where the wind has swept the ground clear.

While eating, the sheep must be alert to such enemies as wolves, bears, coyotes, and mountain lions. Bighorns feed mainly on grasses. They swallow the grass after chewing it only a little. The food is stored in the bighorn's stomach. After eating, the animal brings up a wad of the partly digested food—called a cud. It chews it thoroughly. Then it swallows the cud, so that it can be digested further.

Late in May, many bighorn ewes pick their way along rocky cliffs to high ledges where enemies can't reach them. There the lambs are born. A ewe and her lamb remain by themselves for about a week.

Then they join the other females and young. Bighorn lambs play often. They butt heads, paw the ground, and jump into the air. As the lamb grows, it spends less time with its mother. The ewe pays little attention to her offspring, except to nurse it. She may defend her young, though, especially against golden eagles, which swoop down from the sky.

SHEEP

HEIGHT: 22-51 in (56-130 cm) at the shoulder

WEIGHT: 40-450 lb (18-204 kg)

HABITAT AND RANGE: mountains, forests, and rocky regions in North America, Europe, and Asia; domestic sheep are found in many parts of the world

FOOD: grasses, herbs, leaves, shoots, and twigs

LIFE SPAN: as long as 25 years in the wild

REPRODUCTION: 1 to 4 young after a pregnancy of about 5 or 6 months, depending on species

ORDER: artiodactyls

Bighorns and other wild sheep look very different from the domestic, or tame, sheep raised by people in many parts of the world. Some scientists think that domestic sheep are all descended from one kind of wild sheep—the mouflon (say MOO-flun).

The mouflon is one of the smallest of all the wild sheep. In recent times, it lived only on rocky islands near Italy. By the mid-1800s, the mouflon had nearly died out there because of hunting. But people took some of the animals into forested mountains in France, Austria, and Germany. Though they are still rare in their island homes, thousands of mouflons live in other parts of Europe. Like other wild sheep, mouflons may have summer and winter ranges. But mouflons do not live very high in the mountains. They often stay on lower, forested slopes.

A close relative of the mouflon lives in parts of Asia. The urial (say OOR-ee-ul) is usually larger than the mouflon. All male urials have neck ruffs of long hair. Asian sheep live mainly in rolling country and on high plains, rather than in steep, craggy areas. Most Asian sheep are thinner and have longer legs than the bighorns of North America.

The largest of all wild sheep also lives in the wilderness areas of Asia. The argali (say ARE-guh-lee) of Siberia and Mongolia may measure more than 4 feet (122 cm) tall at the shoulder. It can weigh as much as

▽ *Dall's sheep graze on a mountain slope in Alaska. Because of their slender, widely curved horns, people sometimes call the animals thinhorns.*

White coats dot a grassy ridge as Dall's sheep ▷ *graze in front of Alaska's Mount McKinley, the highest peak in North America.*
▽ *Come on, get up! A Dall's lamb about one week old paws its mother's back. It may want to play.*

△ *Surefooted Dall's ewe leads her lamb across the rocky face of a mountain.*

△ *Crowned with curving horns, a Dall's ram grazes in Alaska. A sheep's horns grow throughout the animal's life. On older rams, the horns may form more than a complete circle.*

Dall's sheep: 36 in (91 cm) tall at the shoulder

German Heath sheep: 27 in (69 cm) tall at the shoulder

Mouflon: 27 in (69 cm) tall at the shoulder

△ Shepherd and his sheep dog herd a flock of Romney ewes through a rolling pasture in New Zealand. Domestic, or tame, sheep like the Romney look very different from their wild relatives. Romney sheep lack horns. People raise the sheep for meat and for their long wool.

△ Long coat of grayish white wool covers an adult German Heath sheep, a centuries-old breed. The black wool of a newborn will change color as the offspring grows older. People raise these sheep for milk, meat, and wool.

◁ Mouflon ram chews his cud in a forest clearing in France. There he finds shelter from winter winds. Mouflons had nearly died out about a hundred years ago. But today they live in many parts of Europe. Some scientists think that all breeds of domestic sheep developed from the mouflon.

512

450 pounds (204 kg). Slightly smaller is the Marco Polo sheep, a kind of argali named for the explorer who traveled from Italy to Asia in the 1200s. Marco Polo sheep are famous for the size of their horns. A horn may measure 6 feet (183 cm) along the curl!

Scientists think sheep were first tamed at least 10,000 years ago. People probably used the animals for meat, hides, and milk. The first domestic sheep had coarse outer coats of straight hair with soft, fine undercoats. Over the centuries, people have developed breeds with coats of fleecy wool. The wool is sheared off, spun into yarn, and woven to make cloth. Sheep are not hurt by the shearing. Their wool grows back in a few months.

Some sheep, like the Merino (say muh-REE-no) sheep, a breed developed in Spain, are known for their very fine wool. They were once so valuable that it was a crime to take a Merino sheep out of Spain. Long-wooled Romney sheep were first developed in England for meat, called lamb or mutton, as well as for wool. Shropshire sheep from England have been bred mainly for meat.

Thin grooves on the deeply ridged horns of a Punjab urial in Pakistan show where the horns stopped growing each fall. This five-and-a-half-year-old ram has a neck ruff of long hair that will fall out after the mating season. Urial ewes have no ruffs. Their horns do not grow as large as the males' horns. Like goats, urials have pointed ears.

Punjab urial: 31 in (79 cm) tall at the shoulder

Shrew

(*say* SHROO)

SHREW

LENGTH OF HEAD AND BODY: **1-12 in (3-30 cm); tail, 1-10 in (3-25 cm)**

WEIGHT: **less than $\frac{1}{15}$ oz-19 oz (2-539 g)**

HABITAT AND RANGE: **almost every kind of habitat throughout most of Asia, Africa, Europe, North America, and northern parts of South America**

FOOD: **insects, worms, snails, and other small animals**

LIFE SPAN: **1 to 4 years in the wild**

REPRODUCTION: **1 to 10 young after a pregnancy of about 2 to 8 weeks, depending on species**

ORDER: **insectivores**

Golden-rumped elephant shrew: 12 in (30 cm) long; tail, 10 in (25 cm)

△ *With its long, pointed snout, a golden-rumped elephant shrew probes for insects in a forest in Kenya. Elephant shrews—distant relatives of the other shrews—spend the day looking for food on the forest floor.*

▽ *Plump and furry, a short-tailed shrew—shown about life size—seems harmless. For its weight, however, the shrew is one of the world's fiercest animals. A short-tailed shrew can quickly kill small prey with its poisonous bite.*

Short-tailed shrew: 3 in (8 cm) long; tail, 1 in (3 cm)

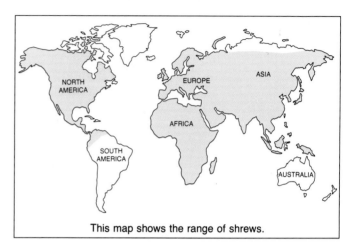

This map shows the range of shrews.

FEROCIOUS AND OFTEN HUNGRY, the shrew is quick to attack small animals, including other shrews. For its size—just inches long—it is one of the fiercest animals in the world. It keeps its active body supplied with energy by feeding almost constantly. Many shrews eat their weight in food every day. A 50-pound (23-kg) child would have to eat about 200 hamburgers to do the same thing!

Sniffing about, a shrew searches for such foods as insects, worms, snails, and other small animals. One kind of shrew, the short-tailed shrew, has poison in its saliva that kills small prey. The animal will even attack and eat a large mouse.

The short-tailed shrew is only one of more than thirty species, or kinds, of shrews found in the United States. In all, more than 200 species of shrews live in many parts of the world.

Most shrews look like sharp-nosed gray or brown mice. But shrews are not rodents. They are relatives of moles. Shrews have soft, dense fur that is darker on their backs than on their bellies. Although a few kinds of shrews have bodies 12 inches (30 cm) long, most of the animals are much smaller. The tiniest shrew—the Etruscan pygmy shrew—is one of the smallest mammals on earth. Shorter than 3 inches (8 cm) from its nose to the tip of its tail, it weighs less than a dime.

Shrews have keen hearing. Some kinds make

Northern water shrew rests on a log before taking a ▷ *dip in a lake in Colorado. A good swimmer, the animal spends most of its life in or near streams and lakes.*

Northern water shrew: 3 in (8 cm) long; tail, 3 in (8 cm)

Common shrew: 3 in (8 cm) long; tail, 2 in (5 cm)

◁ Exploring a bed of moss and grass, a common shrew searches for worms and insects in an English woodland. A sensitive snout leads the animal to its prey.

high-pitched, clicking sounds that human beings cannot hear. When the sounds hit an object, echoes bounce back. From the echoes, shrews can locate food—or obstacles. This process is called echolocation (say ek-oh-low-KAY-shun). Bats and porpoises also use echolocation to hunt and to find their way. You can read about them under their own headings.

Shrews seek shelter in piles of leaves, between rocks, or in underground burrows. Many sleep during the day and come out at night. Others scurry about during both the day and the night, taking short rests from time to time.

Many shrews live alone in areas called territories. Some kinds of elephant shrews—named for their long, flexible snouts—have scent glands at the bases of their tails. They mark their territories with an oily substance from these glands. If one elephant shrew comes into another's territory, the intruder is quickly chased away. Other kinds of shrews claw and bite unwelcome visitors. Usually they do not fight to the death. Instead, the loser surrenders by lying on its back and squealing.

Some kinds of shrews give birth to young about two weeks after mating. These shrews may have several pregnancies a year. Females bear from one to ten offspring at a time, depending on the species.

Follow the leader! A chain of bicolor white-toothed shrews moves like a snake through the woods. Young shrews fall into line behind their mother. Each bites onto the fur of the animal in front of it.

Siamang
The siamang is a kind of gibbon. Read about gibbons on page 222.

Sifaka
The sifaka is a kind of lemur. Read about lemurs on page 324.

Skunk

Stretching for food, a striped skunk in Minnesota licks insects from a stump.

BUSHY TAIL WAVING and front feet stamping, the skunk tries to scare an attacker away. If it does not succeed, it turns around, looks back, and sprays a bitter-smelling mist. Some skunks can spray an area as far as 10 feet (305 cm) away.

The spray—an oily, yellow liquid—is produced by two glands under the skunk's tail. The smell is very strong and can linger for days. The spray may produce a burning feeling in the eyes of the victim. But it causes no permanent damage.

Because of its scent, the small black-and-white skunk has little to fear. It trots boldly through open woodlands and across fields in North America.

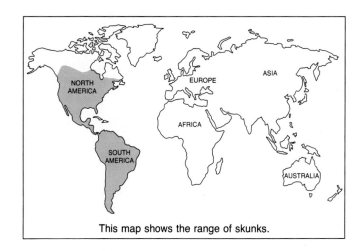

This map shows the range of skunks.

517

Hog-nosed skunk: 16 in (41 cm) long; tail, 9 in (23 cm)

△ Tail in the air, a hog-nosed skunk in Chile looks for small prey to eat. Its broad, bare snout gives the animal its name. The blunt shape of the nose allows the hog-nosed skunk to root easily in the dirt for insects and larvae. It also feeds on rodents.

Spotted skunk waves its ▷ bushy, white-tipped tail in a canyon in Arizona. Although most skunks are marked by broad bands of white fur, this animal's coat has spots and streaks.

Spotted skunk: 11 in (28 cm) long; tail, 7 in (18 cm)

Skunk

Claws gripping the bark, a pygmy spotted skunk ▷ pauses on a log. These rare Mexican skunks move more nimbly than most of their larger relatives.

The hog-nosed skunk also roams South America. Though most skunks are small animals, large meat eaters often avoid them. Foxes, bobcats, coyotes, and owls may attack skunks if there is little other food. But after they have been sprayed once, they probably will have learned their lesson and will stay away. Some scientists think that the skunk's markings serve as a warning signal to other animals.

There are several kinds of skunks. All have black-and-white fur. But the patterns of their coats vary. The striped skunk is recognized by the two white stripes along its back. The hog-nosed skunk often has a white back and tail. The spotted skunk has streaks and spots of white on its body. Its tail is tipped with white.

Skunks take shelter in burrows built by other animals. There they may make a dry, leaf-lined nest. They also live in piles of rocks or in hollow

Pygmy spotted skunk: 8 in (20 cm) long; tail, 5 in (13 cm)

▽ *Spotted skunk takes its warning position. To spray an attacker, it returns to all fours. Before they stand on their forefeet, skunks usually try to frighten enemies away by stamping their feet and waving their tails.*

logs. Sometimes they build nests in old buildings.

When the weather is cold, striped skunks often sleep for a few weeks at a time. Several may huddle in the same den. At other times of the year, skunks usually rest most of the day. At night, they prowl and sniff among leaves and underbrush. They search for insects and larvae, earthworms, eggs, reptiles, and small mammals.

A female skunk bears from two to ten young once a year. Though the newborn are hairless, they already have light and dark markings in their skins. Young skunks may leave their mother after four months. But some sleep in her den all winter.

SKUNK

LENGTH OF HEAD AND BODY: **8-19 in (20-48 cm); tail, 5-15 in (13-38 cm)**

WEIGHT: **7 oz-14 lb (198 g-6 kg)**

HABITAT AND RANGE: **open woodlands, brushy areas, fields, prairies, and deserts in North and South America**

FOOD: **insects and larvae, earthworms, eggs, reptiles, small mammals, fish, fruit, and plants**

LIFE SPAN: **10 years in captivity**

REPRODUCTION: **2 to 10 young after a pregnancy of at least 2 months, depending on species**

ORDER: **carnivores**

Sloth

(*say* SLAWTH)

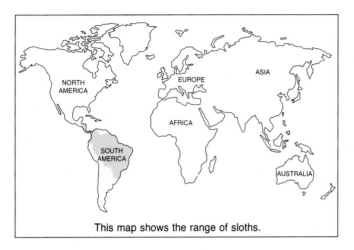

This map shows the range of sloths.

MOTIONLESS FOR HOURS, a sleeping sloth hangs upside down from a branch or curls up in the fork of a tree. The fur of this shaggy animal blends well with the trees of dense forests in Central and South America. Because the climate where it lives is moist and warm, tiny plants called algae often grow on the animal's fur. The plants give a green shimmer to the sloth's grayish brown coat.

This protective coloring makes the sloth difficult to see in the trees and helps hide it from

Three-toed sloth: 22 in (56 cm) long; tail, 2 in (5 cm)

△ *Young three-toed sloth hugs its mother, clinging tightly to her long fur. A sloth offspring rides with its mother until it reaches the age of nine months.*

Under a roof of broad leaves, a shaggy ▷ *three-toed sloth hangs on and stretches out an arm. The number of claws on each front foot makes it easy to distinguish the animal from its two-toed relative.*

◁ *Arm over arm, a three-toed sloth swims in a river in Colombia. By accident, a sloth sometimes tumbles from a tree into the water. This good swimmer may travel a long way before getting out and climbing another tree. On the ground, however, a sloth is the slowest mammal on earth. Its belly rubs the ground as it pulls itself along with its front claws. The animal's weak hind legs and long claws make walking impossible.*

△ *Two-toed sloth makes its way along a tree branch upside down. When it rains, water runs off the sloth's long, coarse fur. Unlike the coats of other animals, a sloth's fur grows from its belly toward its back.*

◁ *From any direction, juicy leaves look good to a two-toed sloth. Because the animal has no front teeth, it must pull the food off with its lips. It chews with side teeth. The leafy diet contains almost all the water a sloth needs.*

enemies. The sloth, about 2 feet (61 cm) long, is the slowest mammal in the world. It cannot outrun the jungle cats that prey on it. It cannot run at all!

When cornered, a sloth tries to defend itself by clawing and nipping. Its heavy, coarse fur and thick skin give it some protection.

Unlike its relatives, the anteater and the armadillo, the sloth is almost helpless on the ground. The sloth's curved claws—useful for climbing—make it hard for the animal to stand. The sloth's legs are too weak to support its weight. To move on the ground, it must lie on its stomach and reach ahead for a claw hold. Then it slowly drags its body forward. In the water, though, the sloth is a good swimmer.

Sloths spend almost all of their lives in the trees. Sloths sleep, eat, and mate high above the ground. Females give birth to young while they hang from tree limbs by their hooklike claws. Sloths have such

firm grips that, even when they die, they sometimes remain attached to branches!

The word *sloth* means "laziness." The animal lives up to its name. A sloth sleeps about 15 hours each day. At night, it travels through the trees in slow motion, feeding on leaves, shoots, and fruit. It rarely drinks water. The sloth gets most of the moisture it needs by eating leaves and licking dew.

The sloth does not hear well. But it has keen senses of smell and touch, which it depends on to find food. The sloth also has very good eyesight. And it can turn its flexible neck far to each side so that it can see in almost every direction.

All sloths have three toes on each hind foot. But the number of toes on their front feet varies. Scientists divide sloths into groups by the number of toes the animals have on each front foot: two-toed sloths and three-toed sloths. A three-toed sloth is also called an ai (say EYE) because of its long, drawn-out call: "ah-eee."

Sloths live alone or occasionally in pairs. A female sloth gives birth to one offspring a year. The newborn sloth has tiny claws, which it uses to climb onto its mother's belly soon after birth. For about a month, it clings to her long hair. Then it begins to move about by itself. When the young sloth is about nine months old, the mother forces it to go off on its own. She nips her offspring whenever it tries to catch a ride with her.

SLOTH

LENGTH OF HEAD AND BODY: 20-25 in (51-64 cm); tail, as long as 3 in (8 cm)

WEIGHT: 9-20 lb (4-9 kg)

HABITAT AND RANGE: forests in parts of Central and South America

FOOD: leaves, shoots, and fruit

LIFE SPAN: as long as 30 years in captivity

REPRODUCTION: 1 young after a pregnancy of about 6 months

ORDER: edentates

Solenodon

(say SO-LEAN-uh-don)

SOLENODON

LENGTH OF HEAD AND BODY: about 12 in (30 cm); tail, about 9 in (23 cm)

WEIGHT: about 2 lb (1 kg)

HABITAT AND RANGE: remote forests and rocky, shrubby areas in parts of the West Indies

FOOD: insects, worms, and small reptiles

LIFE SPAN: about 10 years in captivity

REPRODUCTION: 1 to 3 young after a pregnancy of unknown length

ORDER: insectivores

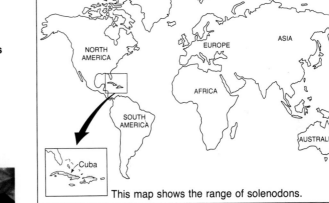

This map shows the range of solenodons.

WITH ITS POINTED SNOUT and long, hairless tail, the solenodon looks like a member of the rat family. But it is not. This animal's relatives include the shrew and the mole. A rare animal, the solenodon lives only in a few remote parts of the West Indies.

As the solenodon moves about, it makes clicking noises in its throat. Echoes that are caused when the sounds hit objects may help the solenodon find its way. It digs into hollow logs with its sharp front claws, looking for insects, worms, and small reptiles. Glands in the solenodon's mouth produce a deadly liquid. The animal's bite poisons its prey.

A female solenodon bears one to three young.

Snout up, a Haitian solenodon sniffs the night air for the scent of prey—insects and other small animals.

Haitian solenodon: 12 in (30 cm) long; tail, 9 in (23 cm)

Springbok

The springbok is a kind of antelope. Read about antelopes on page 52.

Squirrel

(*say* SKWURL)

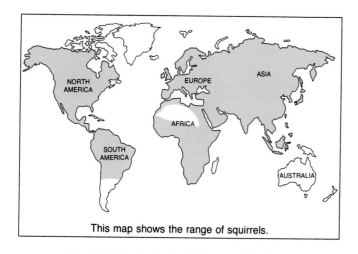

This map shows the range of squirrels.

BRIGHT-EYED AND BUSHY-TAILED, a tree squirrel scampers headfirst down a tree. The furry rodent searches for food on the ground. Suddenly rising on its hind feet, the squirrel looks around. It senses danger and leaps onto a nearby tree, disappearing behind the trunk. Back on a limb near its nest, it flicks its tail, noisily chattering the entire time.

Squirrels live almost everywhere on earth. There are more than 200 species, or kinds. They are found in forests, deserts, mountains, and grass-lands. They range in size from the African pygmy squirrel, only about 5 inches (13 cm) long including its tail, to the Indian giant squirrel, about 3 feet

Harris's antelope ground squirrel: 6 in (15 cm) long; tail, 3 in (8 cm)

△ *White fur rings the eye of a red squirrel eating a plant in Alaska. These small, noisy tree squirrels of North America often chatter loudly in trees.*

(91 cm) long. Though most squirrels have coats of gray or brown, some may be red, black, or white. The Indian giant squirrel has a brightly colored coat of red, black, and pale yellow. The thirteen-lined ground squirrel has stripes and spots.

Despite their differences, all squirrels have front teeth—two upper and two lower—that continue to grow all their lives. Like all rodents, squirrels must gnaw to keep these teeth worn down.

There are three main groups of squirrels: ground squirrels, tree squirrels, and flying squirrels. Ground squirrels rarely climb trees. Some kinds live alone in burrows that they dig. Others, like the California ground squirrel, live together. Their tunnel systems may have several openings and may extend hundreds of feet.

During the day, ground squirrels eat low-growing plants and other food that they find on the

Arctic ground squirrel: 12 in (30 cm) long; tail, 5 in (13 cm)

△ *Nose to cheek, two arctic ground squirrels nuzzle each other. These large squirrels hibernate in their burrows for seven months or more during cold winters in parts of northern North America.*

◁ *Harris's antelope ground squirrel in Arizona perches carefully on the spines of a barrel cactus. It eats the fruit at the center. This kind of squirrel also feeds on other desert plants, seeds, and insects.*

525

Golden-mantled ground squirrel: 7 in (18 cm) long; tail, 4 in (10 cm)

ground—seeds, roots, bulbs, nuts, and leaves. They also eat insects. To catch a grasshopper, a ground squirrel will chase it and pounce. It pins the insect to the ground with its front paws and then bites off its head. A squirrel kills a caterpillar by striking it with the claws on its front feet. All ground squirrels have pouches in their cheeks. They use these to carry food to storerooms in their burrows.

Some kinds of ground squirrels in cold regions spend most of the winter sleeping in their burrows. This sleep is called hibernation (say hye-bur-NAY-shun). When a ground squirrel hibernates, it rolls up in a ball. Its tail curls over its head, and its nose touches its belly. The squirrel lies very still. Its breathing and heart rate slow down, and its temperature drops. It eats little during the cold months, waking up only a few times. It usually lives on fat it

In the Rocky Mountains of Wyoming, a golden- ▷
mantled ground squirrel carries grass for its nest.

Kaibab squirrel: 11 in (28 cm) long; tail, 8 in (20 cm)

△ *With food in its paws, a white-tailed Kaibab squirrel in Arizona sits up for a snack. This kind of tree squirrel has tufts on its ears. It lives only in a small area north of the Grand Canyon.*

▽ *Stripes and spots mark the coat of a thirteen-lined ground squirrel in North Dakota. These animals inhabit the plains and prairies of North America.*

Thirteen-lined ground squirrel: 5 in (13 cm) long;
tail, 4 in (10 cm)

▽ *From a boulder in a mountain meadow in California, a Belding ground squirrel calls out a warning. A fox or a weasel may have come close. At the sound, other ground squirrels nearby will run for cover.*

Belding ground squirrel: 8 in (20 cm) long; tail, 3 in (8 cm)

△ *Northern flying squirrel glides above a snowy slope in the Rocky Mountains. Flying squirrels do not really fly. They can glide, however—as far as 150 feet (46 m)!*
▽ *Spreading front and hind legs, a flying squirrel stretches flaps of skin along the sides of its body. Its flat, furry tail helps the animal steer. When it nears a tree, the squirrel will raise its body and its tail. The skin flaps will act as brakes, slowing the animal as it lands. It will grip the trunk with all four feet.*

has stored in its body. About six months later, when spring comes, the ground squirrel pops out of its burrow. It is much thinner than it was in the fall.

Some ground squirrels in dry areas sleep in summer, too. In deserts of Asia, ground squirrels called large-toothed susliks (say SUH-slicks) go into their burrows in the summer when food is scarce. This summer sleep is called aestivation (say es-tuh-VAY-shun). The antelope ground squirrel, which also lives in deserts, has other ways of surviving high temperatures. If the animal gets too hot, it will stretch out on the ground in a shady spot or go underground where it is cooler.

Like most rodents, ground squirrels have many enemies. Meat eaters such as hawks, foxes, coyotes, and weasels often prey on them. Some ground squirrels may stand on their hind legs to watch for these predators (say PRED-ut-erz), or hunters. If they spot danger, the squirrels give an alarm whistle before running for their underground homes.

Tree squirrels escape from these same kinds of enemies by moving quickly among the trees. Tree

527

Squirrel

Dining upside down, a gray squirrel in Minnesota ▷ *hangs from a tree trunk by its hind claws. This position leaves the animal's forepaws free to hold food.*

squirrels are the animals that people often see scurrying through woods and city parks. With their powerful hind legs, tree squirrels jump easily from branch to branch. Long, bushy tails help them balance as they run along tree limbs.

Sometimes a tree squirrel will run headfirst down a tree. It races jerkily from one side to the other. Sharp, hooklike claws help it cling to the trunk.

Tree squirrels spend most of the time in the trees. But they come to the ground to look for nuts, berries, seeds, and mushrooms to eat. Sometimes a tree squirrel will bury nuts and return to dig them up later. It may not find all of them. Trees often sprout from a squirrel's uneaten meals.

There are many kinds of tree squirrels. North American red squirrels, fox squirrels, chickarees, and Kaibab squirrels live in parts of North America. Another kind of red squirrel lives in Europe and

▽ *Tail laid along its back, a Cape ground squirrel stands and nibbles on food in Namibia, in Africa. White stripes along its sides mark the animal's coarse, brownish hair.*

Cape ground squirrel: 9 in (23 cm) long; tail, 8 in (20 cm)

Gray squirrel: 9 in (23 cm) long; tail, 9 in (23 cm)

Fox squirrel: 12 in (30 cm) long; tail, 10 in (25 cm)

△ *Fox squirrel in Nebraska gnaws an ear of corn. This large North American tree squirrel often lives in city parks as well as in woodlands.*

SQUIRREL

LENGTH OF HEAD AND BODY: 3-18 in (8-46 cm); tail, 2-18 in (5-46 cm)

WEIGHT: ¹/₂ oz-4 lb (14 g-2 kg)

HABITAT AND RANGE: almost every kind of habitat worldwide

FOOD: plants, especially nuts and seeds, and some insects, small birds, and birds' eggs

LIFE SPAN: as long as 15 years in captivity

REPRODUCTION: usually 2 to 8 young after a pregnancy of 3 to 6 weeks, depending on species

ORDER: rodents

528

Eurasian red squirrel: 10 in (25 cm) long; tail, 8 in (20 cm)

△ *Eurasian red squirrel feeds as it perches on a branch of a maple tree in West Germany.*

Asia. The gray squirrel lives in parts of Europe, Africa, and North America.

Flying squirrels spend most of the time among branches, too. They make their homes in hollow trees, in holes made by woodpeckers, and in birds' nests. Flying squirrels usually look for food at night. Their big, bulging eyes help them find the food they eat—nuts, fruit, insects, and even baby birds.

Flying squirrels do not really fly. They glide through the air. To move from one tree to another, a flying squirrel first studies the distance, turning its head from side to side. Pushing off from a limb, the animal stretches out its arms and legs. Flaps of skin extend along the sides of its body and connect its limbs. The animal pushes off and glides to another tree. It steers by moving its legs and tail.

Before it lands, a flying squirrel raises its body and its tail. The skin flaps slow the animal down. The animal's claws grip the surface as it brings all four feet to the tree. A glide can take a flying squirrel 150 feet (46 m) through the air!

Female flying squirrels and tree squirrels give birth in nests of bark, twigs, leaves, and moss. Ground squirrels make their nests in their burrows.

Squirrel

Indian giant squirrel uses its bushy tail to balance as ▷ *it munches fruit. Unlike many other squirrels, these brightly colored rodents do not sit up while eating.*

A squirrel usually bears two to eight blind, hairless young once or twice a year. The offspring stay in the nest one or two months while their mother nurses them. Then they begin to move around outside and to hunt for food.

Squirrels are in the same family as marmots, chipmunks, and prairie dogs. Read about these other animals under their own headings.

Indian giant squirrel: 18 in (46 cm) long; tail, 18 in (46 cm)

Stoat

Stoat is another name for a kind of weasel. Read about weasels on page 564.

Sugar glider

(say SHOOG-er GLY-der*)*

FRUIT, NECTAR, SAP, AND FLOWERS make up most of the sugar glider's diet. This furry, grayish, squirrel-like animal feeds mostly on sweet food. Occasionally, it eats insects. Sugar gliders grunt, chirp, and gurgle as they eat. They also make noise as they move among tree branches in the forests of Australia and New Guinea. Sugar gliders have few enemies, so they don't have to be quiet.

The sugar glider gets to its food by sailing through the air. It extends thin flaps of skin along the sides of its body that connect its wrists and ankles. The sugar glider leaps into the air and glides from one tree to another. It steers by moving its fluffy tail. The animal can travel as far as 150 feet (46 m) in a single glide.

Sugar gliders are members of the phalanger

▽ *Without a magic carpet, a sugar glider in Australia sails through the air. These marsupials extend flaps of skin that connect their limbs. They move from tree to tree, often traveling 150 feet (46 m) in one glide.*

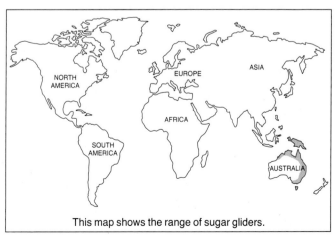

This map shows the range of sugar gliders.

Striped face close to a blossom, a sugar glider licks sweet nectar from flowers of a eucalyptus tree.

family. Like their relatives, sugar gliders are active at night. For this reason, people rarely see the bushy-tailed animals. During the day, groups of sugar gliders curl up in nests of leaves and twigs hidden in holes in trees.

Each nest may shelter two gliders or a family group of a dozen or more animals. Several generations of gliders may share the same nest.

Like all phalangers, sugar gliders are marsupials (say mar-soo-pea-ulz), or pouched mammals. A female glider bears one to three tiny, underdeveloped young. The hairless offspring usually stay in their mother's pouch for several weeks, nursing and growing. Then they begin to climb in and out of the pouch. They go almost everywhere with their mother for a few weeks more, riding on her back.

Read about other phalangers on page 436.

SUGAR GLIDER

LENGTH OF HEAD AND BODY: 5-7 in (13-18 cm); tail, 6-8 in (15-20 cm)

WEIGHT: 3-5 oz (85-142 g)

HABITAT AND RANGE: forests in parts of Australia and New Guinea

FOOD: fruit, nectar, sap, flowers, and insects

LIFE SPAN: about 10 years in captivity

REPRODUCTION: 1 to 3 young after a pregnancy of about 3 weeks

ORDER: marsupials

Suni
The suni is a kind of antelope. Read about antelopes on page 52.

Suricate
(say SOOR-uh-kate*)*

Motionless suricates stand in the sun at the entrance of their burrow in southern Africa.

ON BRIGHT, CLEAR MORNINGS, groups of suricates bask in the sun outside their burrows. As many as 25 of these slender, squirrel-size animals live together in burrows on the plains of southern Africa.

While some of the suricates sit on the ground, others stand on their hind legs and watch the sky for such enemies as eagles and hawks. If a suricate spots danger, it gives a shrill cry of alarm. The others dash for the safety of their burrows.

As they search for food, suricates move slowly with their noses to the ground. They scratch in the dirt for insects and larvae. Sometimes they catch mice and lizards. While suricates hunt, they make purring sounds to stay in contact with each other.

◁ *Female suricate stands up to nurse her offspring. Male suricates also care for the young by guarding them, grooming them, and playing with them.*

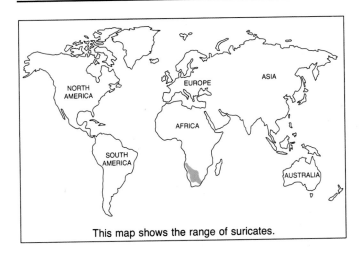

This map shows the range of suricates.

Suricates never wander far from their burrows. But the group often moves from place to place in search of food. The animals may dig new burrows with their sharp claws. Or they may move into burrows dug by ground squirrels. The suricate—often called the meerkat (say MEER-cat)—is a kind of mongoose. Yellow mongooses, close relatives of suricates, often share the same burrows. Read about other mongooses on page 372.

Suricates give birth in their burrows and raise young there. A female suricate may have more than one litter of two to four young every year. Suricates are born blind and nearly hairless. They do not leave the burrow until they are about a month old.

The young play much of the time. They wrestle and nip each other. Older brothers and sisters as well as parents groom, guard, and play with the young. They are fully grown within a year.

Suricate keeps watch on a rock, alert for hawks and ▷ eagles. At first, any flying object—even an airplane—will send a young suricate to cover. After several months, it learns which birds mean danger.

SURICATE

LENGTH OF HEAD AND BODY: 10-14 in (25-36 cm); tail, 7-10 in (18-25 cm)

WEIGHT: about 2 lb (1 kg)

HABITAT AND RANGE: dry plains in southern Africa

FOOD: insects and other small animals and some plants

LIFE SPAN: 10 years in captivity

REPRODUCTION: 2 to 4 young after a pregnancy of about 2½ months

ORDER: carnivores

T

Tahr

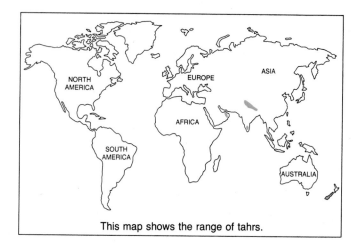

This map shows the range of tahrs.

TAHR

HEIGHT: 24-42 in (61-107 cm) at the shoulder

WEIGHT: 50-240 lb (23-109 kg)

HABITAT AND RANGE: mountainous and hilly regions in parts of Asia and New Zealand

FOOD: grasses, herbs, leaves, shoots, and twigs of shrubs and trees

LIFE SPAN: as long as 21 years in captivity

REPRODUCTION: usually 1 young after a pregnancy of about 6 months

ORDER: artiodactyls

LEAPING NIMBLY among cliffs, tahrs look much like beardless goats. These stocky animals with thick, curving horns are close relatives of wild goats. Read more about goats on page 232.

Rare Nilgiri (say NIL-guh-ree) tahrs are found in the hills and mountains of southern India. Largest of the tahrs, the male Nilgiris measure about 40 inches (102 cm) tall at the shoulder. Except during the mating season, adult males often travel alone or in small groups. Females and young live in herds of five to fifty animals. About six months after mating, a female tahr usually bears one young.

Tahrs eat grasses, herbs, leaves, shoots, and twigs. While the animals rest or feed, they may watch for such enemies as leopards and human hunters. If the tahrs sense danger, they may whistle an alarm and dash to the safety of the cliffs.

There are two other kinds of tahrs. Small Arabian tahrs live only in one dry mountainous area on the Arabian Peninsula. Himalayan tahrs have dark red or brownish black coats and live mainly in the southern parts of the Himalayas. People took them to New Zealand in the early 1900s.

Himalayan tahr: 38 in (97 cm) tall at the shoulder

◁ Long, shaggy mane hangs from the neck of a male Himalayan tahr in Nepal. Several females with shorter coats graze below him. The goatlike animals spend each day feeding on the steep slopes and narrow ledges of mountains. They keep watch for such enemies as snow leopards.

Pointed ears alert, three ▷ Nilgiri tahrs peer watchfully down a steep hillside. The rare animals survive mainly in protected areas of southern India.

Tamandua
The tamandua is a kind of anteater. Read about anteaters on page 48.

Tamarin
The tamarin is a kind of monkey. Read about monkeys on page 376.

Tapir
(*say* TAY-per)

Brazilian tapir: 36 in (91 cm) tall at the shoulder when fully grown

▽ *Wading through a peaceful, marshy pool, a young Brazilian tapir takes an early morning drink after feeding on bushes along the water's edge. Brazilian tapirs spend much of their time in water or mud. Wallowing cools the animals off and rids them of pests—especially ticks. To reach the water, tapirs may slide down wooded hillsides near a pond or a river.*

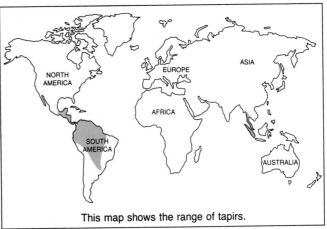

This map shows the range of tapirs.

STANDING MOTIONLESS at the edge of a rain forest, a Brazilian tapir raises its short trunk. The animal sniffs the evening air for the scent of a dangerous enemy—the jaguar. The tapir's white-rimmed ears rotate as it listens for danger. All seems safe. So it steps into a clearing to feed. It nibbles fallen fruit. Using its trunk, it strips leaves and buds from bushes and brings them to its mouth.

Each morning and evening, tapirs visit their feeding and watering places. Although many tapirs may live in the same area, the animals travel alone or in pairs. They trot along tunnel-like paths they have worn through the dense undergrowth during years of use. The tunnels may extend for miles.

Three kinds of tapirs are found in the Western Hemisphere. Both the Brazilian tapir and the Baird's

▽ *Nose and upper lip of a Brazilian tapir form a small trunk (below, left). Like all tapirs, it can stretch its trunk to grasp a branch and bring the leaves to its mouth.*

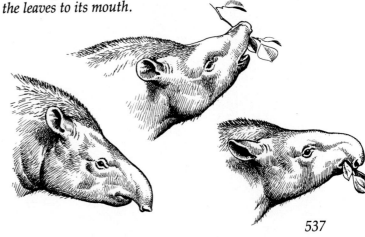

537

△ Sitting next to an adult, a weeks-old tapir looks like a brown watermelon with legs. The patterned coat of the offspring will darken with age.
▽ Brazilian tapir sniffs the forest floor near the Amazon River in Peru. The animal searches for shoots and fallen fruit.

Short, stiff mane bristling, a Brazilian tapir swims easily through the water. It opens its mouth to catch its breath. Or it may be snarling at a pursuer. Now and then, the animal dives to the bottom. There it roots, or digs, for swamp grass and other water plants to eat.

Boldly marked Malay tapir has coloring that helps hide it in the shady forest. The broad white belt around the Malay tapir's middle looks like a blanket. For this reason, people sometimes call the animal the blanket tapir. A tough hide may protect the tapir from the sharp fangs of its enemy, the tiger.

Malay tapir: 42 in (107 cm) tall at the shoulder

tapir make their homes in forests, thickets, and grasslands of South America. The Baird's tapir is also found as far north as Mexico. Near the snow line of the high Andes lives the smallest tapir—the woolly, or mountain, tapir. Its long, wavy coat— much denser than that of other tapirs—protects the animal from chilly temperatures where it lives.

The largest and strongest tapir is the Malay tapir, which weighs as much as 800 pounds (363 kg). It measures about 42 inches (107 cm) tall at the shoulder. This shy animal roams swamps and forests in southeastern Asia. A large patch of light hair makes this kind of tapir look like a black animal with a white blanket tossed over its back. So the animal is sometimes called the blanket tapir.

Scientists think that tapirs have probably lived on earth for about 13 million years. During that time, the animals have changed very little. Once tapirs roamed Europe and North America, but they died out there long ago.

Plump and short-legged, the tapir looks like a large pig. But it is related to the horse and the rhinoceros. Like those animals, the tapir can run quickly, even through tangled vines and thorny bushes.

Tapirs are expert swimmers. To get to the water, they may slide down steep hillsides in dense forests. They also wear down paths on the riverbanks above their swimming holes.

At midday, the animals often wade or wallow in mud. Wallowing helps tapirs rid themselves of ticks. Sometimes tapirs rub their bodies against rocks and tree trunks to scrape off the pests. Or, like dogs, they sit and scratch their chests and front legs with their hind feet.

A female tapir may bear a single offspring at any time of the year. The young tapir has a dark coat patterned with yellow and white stripes and spots. Because of the markings, the offspring is hard to see in the leafy shadows. It keeps its protective coloring for about six months. Within a year, the young tapir is ready to leave its mother and go off on its own.

TAPIR

HEIGHT: 29-42 in (74-107 cm) at the shoulder

WEIGHT: 500-800 lb (227-363 kg)

HABITAT AND RANGE: woodlands, grasslands, and rain forests of southern Mexico, Central and South America, and southeastern Asia

FOOD: water plants, leaves, buds, twigs, and fruit

LIFE SPAN: about 25 years in captivity

REPRODUCTION: usually 1 young after a pregnancy of about 12 or 13 months

ORDER: perissodactyls

Tarsier

(*say* TAR-see-ur)

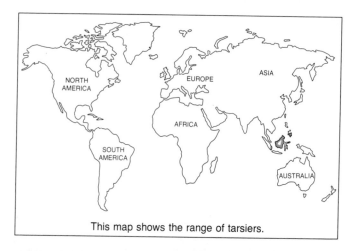

This map shows the range of tarsiers.

HIDDEN BY THE LEAVES of a rain forest, tarsiers move quietly through the darkness. These tiny animals of Indonesia and the Philippines—relatives of apes, monkeys, and human beings—are among the smallest primates in the world.

A tarsier grows only about as large as a chipmunk. But its tail may be twice as long as its body. When it clings to a tree trunk, the animal uses its tail for extra support. The tail also helps a tarsier balance

▽ *Large eyes shining, a western tarsier climbs a tree in a rain forest on Borneo. At night, tarsiers hunt for lizards and insects to eat. They sleep during the day, clinging to tree trunks. Sometimes they hide in hollow trees.*

and steer as it jumps long distances among the trees. The animal can spring almost 2 feet (61 cm) into the air and cover a distance of about 4 feet (122 cm) in a single leap.

Like a frog, a tarsier jumps using its long, powerful back legs. As it leaps, a tarsier jumps backward and twists its body around. It tightly tucks in its arms and legs. It uses its tail to help balance in midair. The tarsier lands feetfirst on a tree trunk with its tail pointed straight up. Flat pads on the ends of its fingers and toes help the animal get firm footing. The tarsier hugs the trunk with all fours.

Tarsiers wake up at sunset and begin to hunt for food. They seek insects, lizards, and other small animals in their homes in the trees. As they move through the forest, tarsiers mark the branches with their urine. By these scent marks, one tarsier announces its presence to others.

Large eyes help the little primate spot its prey in the darkness. The tarsier can turn its head so far around that it can almost see behind itself. Like a bush baby, another small primate, the tarsier can turn each ear in the direction of a sound. It moves its ears almost constantly. The animal's sensitive hearing can detect faint noises in the forest. Read about the bush baby on page 112.

A female tarsier gives birth to a single young. The offspring, born with fur, clings to its mother's belly immediately after birth. Soon the young animal can climb and leap about in the trees.

Long, powerful back legs of a western tarsier help the ▷ *animal climb the slender trunk of a tree. Flat pads on its fingertips and toes give the small primate a sure grip. The tarsier's gray-brown coat helps camouflage, or hide, it in the forest. Huge eyes help it spot prey in dim light.*

TARSIER

LENGTH OF HEAD AND BODY: **3-6 in (8-15 cm); tail, 5-11 in (13-28 cm)**

WEIGHT: **4-5 oz (113-142 g)**

HABITAT AND RANGE: **rain forests in Indonesia and the Philippines**

FOOD: **insects, lizards, small bats, snakes, and small birds**

LIFE SPAN: **at least 8 years in the wild**

REPRODUCTION: **1 young after a pregnancy of unknown length**

ORDER: **primates**

Western tarsier: 5 in (13 cm) long; tail, 10 in (25 cm)

Tasmanian devil

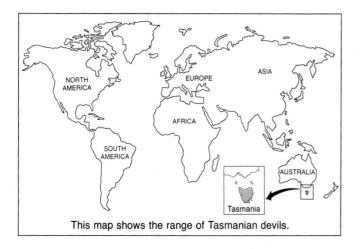

This map shows the range of Tasmanian devils.

TASMANIAN DEVIL

LENGTH OF HEAD AND BODY: 20-31 in (51-79 cm); tail, 9-12 in (23-30 cm)

WEIGHT: 10-20 lb (5-9 kg)

HABITAT AND RANGE: brushy areas on Tasmania

FOOD: small mammals, birds, reptiles, and dead animals

LIFE SPAN: about 8 years in the wild

REPRODUCTION: 3 or 4 young after a pregnancy of about 1 month

ORDER: marsupials

WITH ITS MUSCULAR JAWS and large teeth, the Tasmanian devil snaps up anything it can catch. This meat-eating marsupial (say mar-soo-pea-ul), or pouched mammal, even eats poisonous snakes. It also feeds on dead animals that it finds.

Because of its heavy head and stocky body, a Tasmanian devil looks a little like a black bear cub. But a Tasmanian devil has a hairy tail about 10 inches (25 cm) long. The animal lives only on the island of Tasmania, a part of Australia.

Tasmanian devils hunt at night. With their tails off the ground, they run awkwardly through thick underbrush. During the day, the animals sleep in caves, in hollow logs, or among rocks.

Except during the mating season, Tasmanian devils live alone. If two meet, they scream, snort, spit, and snarl at each other. But the animals rarely attack one another. One of them usually runs away.

Like other marsupials, female Tasmanian devils give birth to three or four tiny, underdeveloped young. Immediately after birth, the raisin-size offspring crawl into the pouch on their mother's belly. They remain there for about four months, nursing and growing larger.

Tasmanian devil's ears turn red with alarm as it stands ready to defend its home among tall grasses. Once found in other parts of Australia, these pouched mammals now live only on the island of Tasmania.

Tenrec

(*say* TEN-wreck)

Sniffing the ground, a tailless tenrec hunts in the ▷ daylight. Usually these animals sleep all day in burrows. ▽ Short, stiff spines stand up on the back of a streaked tenrec. A crest of longer spines grows on its head.

Streaked tenrec: 7 in (18 cm) long

Tailless tenrec: 12 in (30 cm) long

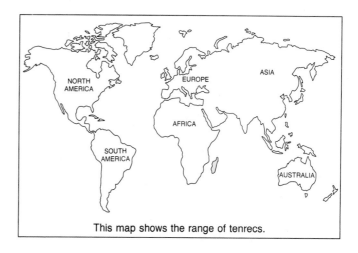

Madagascar hedgehog tenrec: 6 in (15 cm) long

FOR MORE THAN TEN MILLION YEARS, scientists think, tenrecs have lived on Madagascar, an island off the coast of Africa. There these relatives of hedgehogs have developed in many ways. Some tenrecs look like hedgehogs. Others resemble shrews or moles. The tailless tenrec, the largest of the tenrecs, grows as big as a rabbit. The small streaked tenrec fits easily in a person's hand.

During the day, tenrecs rest in burrows. At night, the animals scurry across grasslands and through forests. There they look for insects, earthworms, mice, small reptiles, and roots to eat.

Most female tenrecs bear more than ten tiny offspring two months after mating. But a tailless tenrec may bear more than 25 young in a litter—more young at one time than any other mammal!

△ *Thick coat of spines covers the head and body of a Madagascar hedgehog tenrec. Like its relative the hedgehog, this kind of tenrec rolls into a prickly ball when threatened. Tenrecs live only on Madagascar and a few nearby islands. They eat mainly insects and earthworms.*

TENREC

LENGTH OF HEAD AND BODY: **3-15 in (8-38 cm); tail, as long as 6 in (15 cm)**

WEIGHT: **1 oz-5 lb (28 g-2 kg)**

HABITAT AND RANGE: **grasslands and forests of Madagascar and nearby islands**

FOOD: **insects, earthworms, mice, small reptiles, and roots**

LIFE SPAN: **about 4 years in captivity**

REPRODUCTION: **1 to 25 young after a pregnancy of about 2 months, depending on species**

ORDER: **insectivores**

This map shows the range of tenrecs.

543

Tiger

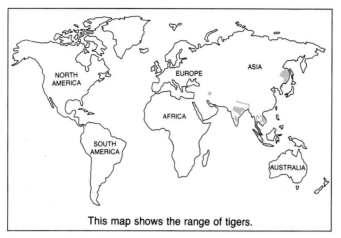

This map shows the range of tigers.

TIGER

LENGTH OF HEAD AND BODY: **5-6 ft (152-183 cm); tail, 2-3 ft (61-91 cm)**

WEIGHT: **240-500 lb (109-227 kg)**

HABITAT AND RANGE: **forests, wooded hillsides, and swamps in parts of Asia**

FOOD: **deer, wild cattle, antelopes, and smaller animals**

LIFE SPAN: **as long as 20 years in captivity**

REPRODUCTION: **1 to 6 young after a pregnancy of about 3½ months**

ORDER: **carnivores**

BEAUTY, MYSTERY, AND STRENGTH are all qualities for which the tiger has been admired and feared. In some Asian myths, the tiger—with its striking pattern of stripes—is the king of beasts. But in most stories, the tiger is a demon. Because of the animal's reputation as a dangerous foe, those who hunted the tiger often were respected for their bravery. Tiger hunting became a popular sport.

A century ago, thousands of tigers roamed much of Asia. Tigers lived wherever water, prey, and places to hide were plentiful. They prowled rain forests, wooded hillsides, and swamps in many parts of Asia. Today they survive only in parts of that range. On the islands of Java and Bali, the tiger is probably extinct. In the forested areas of India, perhaps 2,000 animals remain. The tiger is now protected by law. But the habitats in which tigers live still are being destroyed.

Partly hidden by tall grass and shadows, a Bengal tiger looks across a field in India. No two tigers have the same pattern of stripes on their coats.

Tiger

The tiger is the largest member of the cat family, and the Siberian tiger is the biggest tiger of all. It usually measures about 9 feet (274 cm) long from its nose to the tip of its tail. In winter, its light orange coat is long and thick. Tigers that live farther to the south are smaller. Some are dark in color with heavy black stripes on a reddish orange background. A few tigers are white with brown stripes, but there is no record of any black tigers.

During the day, the tiger rests in the shade. It may lie in a quiet pool of water to escape the heat. At dusk, the tiger begins to hunt for food. Tigers usually prey on deer, wild cattle called gaur, and wild pigs. But if the tiger is hungry and cannot find large

Indian tiger rubs its head against a tree trunk, leaving ▷ its scent. Tigers live and hunt alone. They communicate by scent marks and sounds as well as by spraying urine. ▽ At the edge of a stream, a tiger rests with its hindquarters in the water and its front legs stretched out in the grass. Tigers, the largest of all the cats, spend time in the water to cool off in the daytime heat.

prey, it will eat any kind of meat. A few tigers have become man-eaters. Man-eaters are usually sick or wounded animals that cannot hunt their normal prey. Generally, tigers avoid people.

As it hunts, a tiger uses its keen eyesight and hearing. A tiger cannot run great distances. Instead, it stalks until it is close to its prey. Then it rushes and pulls the prey to the ground with its teeth and claws. It strangles a larger animal by biting the throat, and it breaks the necks of smaller ones.

A tiger may camp near its kill for several days, until it has eaten all the meat and most of the bones. It feeds for a while, grooms itself, takes a drink of water, rests, and then feeds some more. The cat

◁ *Three ten-month-old cubs crowd close to their mother as she rests in a pool. Two of the cubs prepare to greet each other by rubbing their heads together. Offspring usually stay with their mother for about two years.*

▽ *Male tiger charges through a pond in India. Strong swimmers, tigers may cross a large river to find prey.*

Tiger

usually eats about 12 pounds (5 kg) at a time, but it may eat as much as 60 pounds (27 kg) in one night.

Tigers usually live and hunt alone. But when food is plentiful, several animals may gather at a kill. They often stay together for short periods. Tigers keep in touch by scents and sounds. They spray

Widely varying climates provide homes for the tiger. A male and a female Sumatran tiger (right) peer through dense bushes. In Sumatra, tigers often live in swamps. On a mountain slope, two Siberian tigers (below) pause in the snow. People have hunted the big cats for their strikingly marked fur and for sport. And they have settled the wild areas where the animals live. Sixty years ago, tens of thousands of tigers roamed much of Asia. Today fewer than 5,000 survive in the wild.

bushes and trees with urine. They roar, moan, and grunt. If two tigers meet, they may greet each other by rubbing heads and making a puffing sound.

A female tiger usually gives birth every two or three years, about three and a half months after mating. Her litter may include two or three cubs the size of large kittens. Cubs nurse for about six months.

But their mother begins to take them to a kill when they are only two months old. Gradually, the cubs spend more time on their own. Males usually can hunt by themselves earlier than females. The cubs become independent within two years.

Read about other wild cats in the entries on bobcat, cat, cheetah, jaguar, leopard, lion, and lynx.

Topi
The topi is a kind of antelope. Read about antelopes on page 52.

Tuco-tuco

(*say* TOO-koh-TOO-koh)

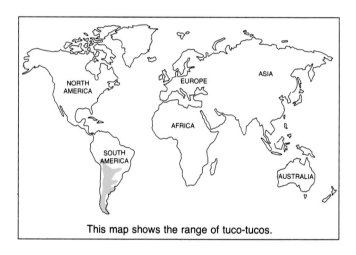

This map shows the range of tuco-tucos.

LIKE THE STRIKING OF A HAMMER, the calls of a tuco-tuco ring out from the animal's underground home. The sounds it makes give this rodent its name. The noises may warn of danger or help the animal claim its territory.

Tuco-tucos look somewhat like pocket gophers. They live in forests and on grasslands and plains in parts of South America. The rodents spend most of the time in burrows they have dug with their claws and large front teeth.

A tuco-tuco may peer out of its burrow to look for enemies. And if a fox, skunk, wild cat, or hawk approaches the animal darts back in. The tuco-tuco rarely leaves its home, even to find food. From inside the burrow, it can pull down plants by the roots. In her burrow, a female bears from one to six young each year.

△ *Tuco-tuco grasps a blade of grass. Usually these stocky rodents of South America stay underground. There they can pull down their food by the roots.*

TUCO-TUCO

LENGTH OF HEAD AND BODY: 6-10 in (15-25 cm); tail, 2-4 in (5-10 cm)

WEIGHT: 4-24 oz (113-680 g)

HABITAT AND RANGE: forests, grasslands, and plains in parts of South America

FOOD: roots, bulbs, stems, and grasses

LIFE SPAN: probably less than 3 years in the wild

REPRODUCTION: 1 to 6 young after a pregnancy of about 3 months

ORDER: rodents

U

Uakari
The uakari is a kind of monkey. Read about monkeys on page 376.

V

Vervet
The vervet is a kind of monkey. Find out about monkeys on page 376.

Vicuña
The vicuña is a relative of the llama. Read about both animals on page 342.

Vizcacha

(*say* vis-KOTCH-uh)

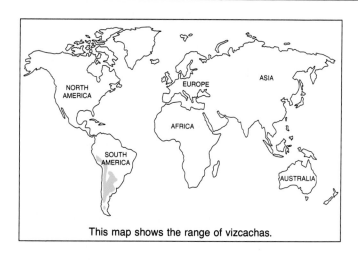

This map shows the range of vizcachas.

VIZCACHA
LENGTH OF HEAD AND BODY: 12-26 in (30-66 cm); tail, 6-15 in (15-38 cm)
WEIGHT: 2-18 lb (1-8 kg)
HABITAT AND RANGE: rocky, mountainous regions and grassy plains in parts of South America
FOOD: plants such as grasses and mosses
LIFE SPAN: as long as 19 years in captivity
REPRODUCTION: 1 or 2 young after a pregnancy of 4 or 5 months, depending on species
ORDER: rodents

TWO RODENTS OF SOUTH AMERICA share the same common name, though they differ in many ways. High in the Andes, the mountain vizcacha leaps gracefully among rocks. The plains vizcacha runs across the grasslands of southern South America.

The mountain vizcacha's body measures about 13 inches (33 cm) long. Its gray or brown fur is thick, soft, and short. A black stripe often runs down its back. The animal has long ears and a tail tipped with black or reddish brown hairs.

During the day, groups of mountain vizcachas feed on grasses and other plants. They also perch on rocks, basking in the sun or grooming their fur.

Mountain vizcachas live in colonies that number as many as several hundred animals. When alarmed, they signal to one another with whistling calls. Then they dash for shelter among the rocks.

The mountain vizcacha's larger relative, the plains vizcacha, has a stockier body that usually measures about 24 inches (61 cm) long. The rodent has black whiskers and black-and-white stripes on its face. Its coarse coat ranges from gray to light brown, and the animal has a white belly.

Plains vizcachas dig networks of burrows with many entrances. As many as thirty plains vizcachas may share a burrow. But the animals often have guests. Lizards, snakes, toads, foxes, and burrowing owls also may live in the underground homes.

At dawn and at dusk, plains vizcachas look for grasses, roots, stems, and seeds to eat. They also pick up unusual objects to carry home. On top of

Mountain vizcacha: 13 in (33 cm) long; tail, 11 in (28 cm)

Plains vizcacha: 24 in (61 cm) long; tail, 7 in (18 cm)

△ *Rabbitlike rodent—except for its tail—a mountain vizcacha basks in sunshine near its home in the Andes of Peru. Agile jumpers, mountain vizcachas live in rock crevices, but they feed in open grassy areas. These vizcachas live in colonies that may number several hundred animals.*

◁ *Plains vizcacha stops for a snack on an open grassland. Larger than its mountain relative, the plains vizcacha has a black-and-white striped face. A junk collector, it picks up bones, stones, and other objects that it finds. Then it piles them on top of its burrow.*

their burrows, these junk collectors heap bones, stones, branches, lumps of earth—even jewelry that people have lost nearby. The pile on a single burrow is often enough to fill a wheelbarrow! Why the animals collect these objects remains unknown.

Vizcachas, like all rodents, have front teeth that grow throughout their lives. The animals must gnaw to keep their teeth from growing too long.

Female vizcachas give birth once or twice a year, four or five months after mating. A female mountain vizcacha usually bears a single young. A plains vizcacha has a set of twins.

Vole

AMONG THE MOST NUMEROUS mammals on earth, voles live in parts of North America, Europe, Asia, and Africa. The stocky, grayish brown vole looks much like its relative the mouse. Voles have shorter tails and ears than mice do, however. There are nearly 200 species, or kinds, of voles.

Voles live in many habitats—forests, grasslands, marshes, and mountains, as well as in orchards and in gardens. Voles eat whatever plants they can find. They nibble on stems, leaves, roots,

◁ *Moving through grass and clover, a bank vole in West Germany watches cautiously for such enemies as snakes, hawks, owls, cats, dogs, and weasels.*

Bank vole: 4 in (10 cm) long; tail, 2 in (5 cm)

▽ *Water vole nibbles a willow leaf at the edge of a small stream in England. Like muskrats, their relatives, water voles swim well, both on the surface and underwater.*

Water vole: 8 in (20 cm) long; tail, 4 in (10 cm)

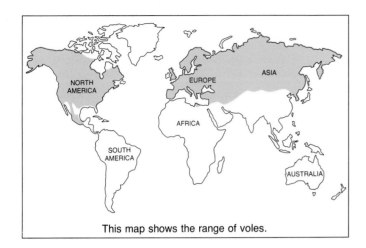

This map shows the range of voles.

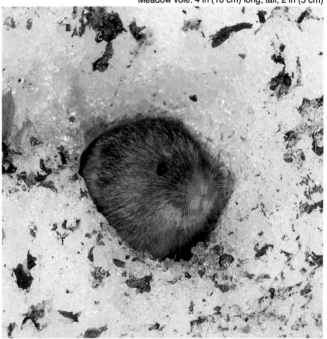

Meadow vole: 4 in (10 cm) long; tail, 2 in (5 cm)

flowers, seeds, grasses, and bark. Some voles, such as the meadow vole, can eat almost their own weight in food in a day—nearly 3 ounces (85 g).

Many voles live in nests of shredded plants under logs or rocks or in trees. Other voles dig burrows. The rodents usually connect their nests to feeding grounds with runways, or networks of paths, through the grass. They keep the paths clear by running back and forth and by nibbling on the plants in their way. In the winter, some voles make their runways under snow.

Voles reproduce often. A female vole can bear several litters a year with as many as eight young in each litter. Blind and helpless, newborn voles weigh less than a quarter of an ounce (7 g). They usually stay in the nest less than one month.

Sometimes the vole population grows too large. The small rodents can destroy entire fields as they try to find enough to eat. Usually, however, the natural enemies of voles keep them from becoming too numerous. Snakes, owls, hawks, cats, dogs, weasels, and many other animals prey on voles.

VOLE

LENGTH OF HEAD AND BODY: 3-8 in (8-20 cm); tail, as long as 4 in (10 cm)

WEIGHT: $^1/_2$ oz-7 oz (14-198 g)

HABITAT AND RANGE: many kinds of habitats in parts of North America, Europe, Asia, and a small area in Africa

FOOD: plants, especially grasses

LIFE SPAN: about 1 year in the wild

REPRODUCTION: 1 to 8 young after a pregnancy of 3 or 4 weeks

ORDER: rodents

Short-tailed vole: 5 in (13 cm) long; tail, about 1 in (3 cm)

Wallaby (say WOLL-uh-bee)

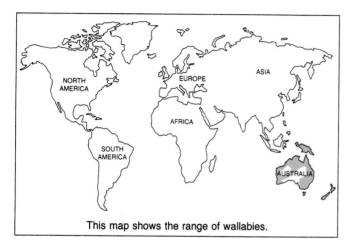

This map shows the range of wallabies.

Red-necked wallaby and her young, called a joey, share ▷ a meal. They carefully strip a thin branch of its leaves. ▽ Miniature marsupial, an unadorned rock wallaby grips the surface of a rock with its rough, thickly padded hind feet. These surefooted jumpers can leap as far as 12 feet (4 m) to get from one rock to another.

Unadorned rock wallaby: 24 in (61 cm) long; tail, 20 in (51 cm)

554

Red-necked wallaby: 35 in (89 cm) long; tail, 29 in (74 cm)

NOT ALL KANGAROOS are called wallabies, but many are. Wallabies are small and medium-size members of the kangaroo family. More than thirty kinds of wallabies live in parts of Australia, New Guinea, and nearby islands. People also took them to New Zealand in the 1800s.

The largest wallaby is almost 6 feet (183 cm) long from its nose to the tip of its tail. The smallest wallaby is about the size of a rabbit. Read about kangaroos on page 310. Find out about the quokka, a small wallaby, on page 470.

The coats of wallabies come in a wide range of colors. Their fur may be gray, brown, red, or nearly black. Some wallabies have patches of bright yellow or orange fur. A light orange ring encircles each eye of the spectacled hare wallaby. The tail of the ring-tailed rock wallaby is ringed with bands of white, brown, and pale yellow.

Red-legged pademelon: 25 in (64 cm) long; tail, 16 in (41 cm)

△ *Red-legged pademelon, a kind of scrub wallaby, searches the forest floor for leaves and grasses to eat. When feeding, the animal moves at a leisurely pace on all fours. If startled, however, it hops away on its hind legs. The pademelon always follows the same paths, or runways, through the woods.*

Sightseeing from a safe spot, a joey peeks through tall ▷ grass from its mother's pouch. The young sandy wallaby may leave the pouch to graze. But it returns at the slightest hint of danger.

Wallabies are usually found in shrubby areas or in rocky regions. The smaller shrub wallabies are often called scrub wallabies or pademelons (say PAD-ee-mel-unz). These animals seek cover among the tangled plants of swamps, thickets, and forests. Larger wallabies are often called brush wallabies. They live among high grasses and tall shrubs.

Rock wallabies inhabit rocky, hilly regions. They find shelter in caves or between rocks during the day. There they are shielded from the hot sun. They also are hidden from such enemies as the wild dogs called dingoes.

In the cooler temperatures of evening, wallabies leave their shelters and look for food. Wallabies feed mainly on plants. They need to drink very little water. When water is scarce, they get most of the moisture they need from the food they eat. One kind of wallaby, the tammar (say TAM-ur), will even drink salt water when fresh water is not available.

When grazing, wallabies move slowly about on all fours. If startled, however, they hop swiftly away on their strong hind legs. Hare wallabies, named for their harelike size and movements, can jump higher than a person's head. If discovered by an enemy, a hare wallaby escapes with twists and turns. Leaping away, it leaves its attacker behind.

Surefooted rock wallabies hop nimbly about Australia's steep, rocky hills. Thick pads of rough skin on the animals' hind feet give them a good grip on the uneven terrain.

Wallabies use their tails for balance as they leap. Most have strong tails that support them when they sit. The most unusual tail among wallabies belongs to the nail-tailed wallaby. This animal has a horny spike at the tip of its long tail.

Wallabies, like all kangaroos and many other Australian animals, are marsupials (say mar-soo-pea-ulz), or pouched mammals. A few weeks after mating, a female wallaby gives birth to one tiny, underdeveloped offspring, called a joey. Blind and helpless, the newborn immediately crawls into its mother's pouch. There it remains, nursing for several months, until it is big enough to digest solid food.

Even when it begins to leave the pouch to graze, the joey stays close to its mother. She continues to nurse and to protect the young wallaby, though, until it can take care of itself.

WALLABY

LENGTH OF HEAD AND BODY: 12-41 in (30-104 cm); tail, 10-29 in (25-74 cm)

WEIGHT: 4-53 lb (2-24 kg)

HABITAT AND RANGE: shrubby, rocky areas of Australia, New Guinea, neighboring islands, and New Zealand

FOOD: mostly grasses and other plants

LIFE SPAN: as long as 15 years in captivity

REPRODUCTION: usually 1 young after a pregnancy of about 1 month

ORDER: marsupials

△ *Heads pulled back, two male sandy wallabies challenge one another (above, left) for a female. With the claws of the front paws, each slashes at the other's body (above, center). One wallaby aims a kick at his opponent's belly (above, right). Such contests rarely end in serious injury.*

Walrus

A BRISTLY MUSTACHE and long ivory tusks make the walrus look like a storybook creature. The animal is real, however. Walruses are relatives of seals and sea lions. Like these animals, walruses live mostly in the water. They are found in the Far North—in the Arctic Ocean and nearby waters. Those that live off Greenland and eastern Canada are usually smaller than the others. In the northern oceans, the water is cold, and huge chunks of ice float on the surface. The floating ice provides walruses with places to bear young and to rest.

◁ *Under the sharply peaked mountains of an island off Alaska, thousands of Pacific walruses bask in the sun. These older males spend the summer in this area. Females and young males head farther north.*

A walrus's tusks are important for survival. Both male and female walruses have tusks. These long teeth keep growing throughout a walrus's life. Some may reach a length of 30 inches (76 cm).

Male walruses—called bulls—often fight with their tusks to determine which animal is stronger and more important. At first, they raise their heads and show off their tusks. Sometimes that is enough to make a walrus with smaller tusks move out of the way. If two walruses do fight, they usually just poke and jab. They rarely hurt one another very badly. The skin of a walrus is thick, and its neck is covered with tough lumps. Usually the largest bull—which often has the longest tusks—wins. During the mating season, injuries are more common. Bulls fight each other in earnest.

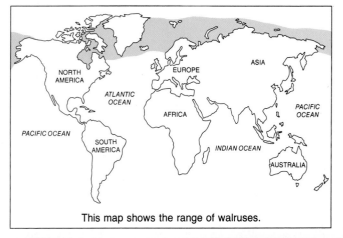

This map shows the range of walruses.

△ *Blast away! A Pacific walrus clears its nose of water. Like all sea mammals, walruses must come to the surface to breathe. They can dive as deep as 260 feet (79 m) and remain underwater for more than ten minutes.*

◁ *Long ivory tusks cast shadows on the thick hides of two Pacific walruses. Both males and females grow tusks, which may reach 30 inches (76 cm) long. With these teeth, the animals can get a hold on pack ice and pull their bulky bodies out of the water.*

Pacific walrus: 12 ft (4 m) long

Sometimes their tusks break off during battles.

The animals also poke each other to keep the best spots on a beach or on a piece of ice. Occasionally, walruses fight off such enemies as polar bears with their tusks. Walruses also use their tusks to help pull their bulky bodies out of the water. A walrus sticks its head up and hooks its tusks on a chunk of ice. Then it hauls itself up.

A walrus uses its mustache when it roots, or digs, for food. The animal dives to the ocean floor and feels in the sandy bottom with its stiff, sensitive whiskers. When it finds snails or clams, it scoops up the shellfish with tongue and lips. It sucks out the meat and leaves the shells on the bottom. Walruses also eat worms, crabs, and shrimps—as much as 100 pounds (45 kg) of food a day. That adds up to about

▽ *Jammed together for warmth, walruses huddle on a rocky beach. Sometimes one walrus will poke another with its tusks. This signal means "move over and make room."*

800 large clams or as many as 10,000 smaller ones!

A thick layer of fat—called blubber—helps protect walruses from the cold. Blubber makes up about one-third of a walrus's total weight. In the winter, when its layer of fat is as thick as 6 inches (15 cm), a walrus may weigh 2 tons (1,814 kg). Despite its weight, the walrus is graceful in the water. To propel itself, it moves its rear flippers from side to side. It can swim about 5 miles (8 km) an hour.

On land, the walrus walks awkwardly. Like many seals, it can use its flippers as legs out of the water. It turns its rear flippers forward and walks along on all fours. The rough bottoms of its flippers prevent it from slipping.

Reddish brown is the usual color of a walrus's hide. But when the animal comes out of the water after a cold swim, it often appears pinkish white. That's because its blood has moved away from the skin surface to the tissues inside its body. This helps the walrus keep its body heat up. When the animal

Walrus

comes ashore, the blood moves back to the skin. This cools the animal. After a few minutes, a walrus's skin returns to its reddish brown color.

Walruses live in herds of as many as several thousand animals. They hunt and rest together. When they come out of the water, they may remain on land or on ice for nearly two days at a stretch. The animals sleep most of the time. They sprawl out, warming themselves in the sun and making deep grunting noises. Polar bears sometimes charge at resting herds. They try to catch calves or slower walruses before they can escape into the water.

As the seasons change, walruses move to other feeding grounds. In the spring, large herds of walruses move north. Some older bull walruses do not make the trip, however. They stay in the same area

△ *Sunlight reaches into the water, streaking two walruses. A walrus moves through the water by sweeping its rear flippers from side to side.*

△ *Surrounded by blue sea, a walrus herd crowds onto huge chunks of ice in the Arctic Ocean. In the small picture, a female walrus nuzzles her months-old calf.*

year round. When the northern waters begin to freeze in the fall, the walruses that did migrate, or travel, swim south again. Walruses are strong swimmers. They can remain in the water for days at a time, although they sometimes catch rides on drifting chunks of ice.

Walruses even sleep in the water. Males and females may doze beneath the surface, or they may hold their heads above the water so they can breathe. An adult male walrus has two pouches in his throat. He can fill the pouches with air until they are as big as beach balls. The pouches help the animal stay afloat. Females do not have these pouches. Male walruses also use their pouches to make deep, ringing sounds. The noises may help attract females during the mating season.

Walruses mate in the water, and the female usually bears a single calf in May. Like seals and sea lions, walruses come onto land or ice to give birth. The newborn calves weigh as little as 85 pounds (39 kg). Short, grayish brown hair covers their wrinkled skin. At first, newborn walruses stay out of the

---

WALRUS

LENGTH OF HEAD AND BODY: 6$\frac{1}{2}$-12 ft (198 cm-4 m)

WEIGHT: 1,650 lb-2 t (748-1,814 kg)

HABITAT AND RANGE: Arctic Ocean and nearby waters

FOOD: clams, snails, crabs, and worms

LIFE SPAN: 40 years in the wild

REPRODUCTION: 1 young after a pregnancy of 12 months

ORDER: pinnipeds

water, but they can swim to escape such enemies as polar bears. At about three months of age, their tusks begin to grow.

A mother walrus watches over her calf carefully. She protects it from icy winds with her body. Later, she may carry it in the water between her flippers. Or it may perch on her back when she swims near the surface. The offspring remains with its mother for about two years. Occasionally, a male calf may stay as long as four years. Then it leaves to join a herd of other young males.

The Eskimo have hunted walruses for hundreds of years. Harpoon heads and other tools were made from walrus bones and tusks. The walrus's stiff whiskers were used for toothpicks. In some places today, oil made from the blubber is still burned for heat and light. And the Eskimo eat walrus meat and cover their boats with walrus hide.

Waterbuck
The waterbuck is a kind of antelope. Read about antelopes on page 52.

Weasel

(*say* WEE-zul)

TIRELESS HUNTERS, weasels bound and zigzag across brushy ground after rodents and small birds. They even dart into burrows after ground squirrels. With its slender body and short legs, a weasel can easily squeeze into underground dens where it can kill its prey. Larger weasels sometimes hunt rabbits. A weasel may kill more food than it needs. It stores the food in a safe place and returns to it later.

Weasels live in meadows, forests, and grasslands in many parts of the world. They take shelter in burrows and rock crevices or among tree roots. Of about ten kinds of weasels, only the long-tailed weasel, the short-tailed weasel, and the least weasel are found in North America.

The weasel has glands that produce a strong-smelling substance. Scientists think that the animal leaves this and its waste in its hunting area. The scent tells other weasels to hunt elsewhere.

◁ *Standing beside a ground squirrel's burrow, a long-tailed weasel in Nebraska listens for prey. The animal keeps its brownish coloring all summer.*

Long-tailed weasel: 12 in (30 cm) long; tail, 5 in (13 cm)

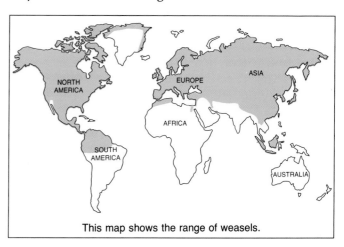

This map shows the range of weasels.

564

Long-tailed weasel in Canada climbs among branches. The coats of many weasels change color with the season. In fall, new, lighter hair gradually grows in. By winter, the weasel's coat is white.

WEASEL

LENGTH OF HEAD AND BODY: 5-16 in (13-41 cm); tail, as long as 7 in (18 cm)

WEIGHT: less than 2-11 oz (57-312 g)

HABITAT AND RANGE: meadows, forests, and grasslands in many parts of the world

FOOD: small mammals, birds, berries, and insects

LIFE SPAN: about 10 years in captivity

REPRODUCTION: 3 to 13 young after a pregnancy of 1 to 12 months, depending on species

ORDER: carnivores

Short-tailed weasel: 10 in (25 cm) long; tail, 3 in (8 cm)

Most weasels have brownish fur with lighter underfur. In northern parts of North America, Europe, and Asia, however, many weasels grow a white coat by winter. They are hard to see in snowy country. The tail tips of some weasels stay black. Enemies such as owls and hawks sometimes swoop at the black spot of tail fur and miss the weasel's body.

The short-tailed weasel is called an ermine when it has its winter fur. In Europe, it is called a stoat when it has its brown summer coat. The white fur of other weasels is also known as ermine.

A female weasel bears from 3 to 13 young once or twice a year. After about four months, the offspring begin to look for their own hunting areas.

◁ *Feeding on the leg of a snowshoe hare, a short-tailed weasel finishes the remains of a lynx's kill.*

▽ *Protective winter coloring almost hides a short-tailed weasel bounding across a snowy mountainside in Switzerland. Hawks or owls may swoop at the black tip of a weasel's tail—and miss the body of their prey!*

△ *Eyes bright, a short-tailed weasel in Alaska pops out of an arctic ground squirrel's burrow. Weasels hunt whenever they are hungry. They may mark their hunting areas with scent.*

Whale

WHAT WEIGHS more than thirty elephants and lives in the sea? The female blue whale—the largest animal in the world! It weighs 200 tons (181 t) or more and grows as long as 100 feet (30 m). Even a baby blue whale is big. Two tons (1,814 kg) at birth, a blue whale calf gains 200 pounds (91 kg) a day until it stops nursing at about one year of age.

People once thought that whales were fish. Whales look somewhat like fish. And they live in water—in all the oceans of the world. But whales actually are large sea mammals. Large whales are often called great whales. Some kinds of smaller toothed whales are called porpoises or dolphins. Find out about them on page 452.

Accompanied by her calf, a female humpback whale (above, left) swims in breeding grounds off Hawaii. Mothers and young often seem to show affection by patting each other with their flippers. A female whale nurses her young by squirting milk into its mouth. The calf stays close to her for nearly a year, drinking her milk and growing larger. The humpback calf will not reach its full size—more than 45 feet (14 m) long—for about ten years.

◁ Air bubbles stream from the blowhole of a humpback whale. The animal may be making sounds—grunts, moans, bellows, or whistles. Humpbacks also sing. Again and again they repeat an eerie melody.

Humpback mother and calf perform a graceful water ▷ ballet as they swim. They swing their broad flukes, or tail fins, up and down to push themselves forward. To steer, they use long, winglike flippers.

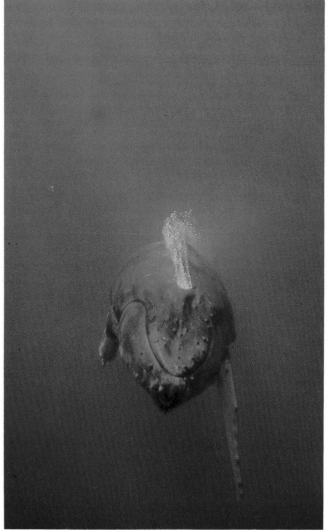

Humpback whale: 45 ft (14 m) long

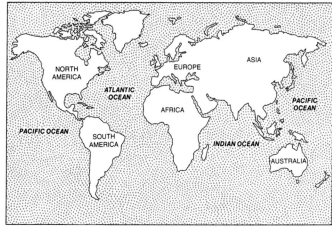

This map shows the range of whales.

Enormous acrobat, a humpback whale leaps skyward. It hurls its body into the air and splashes back into the sea with a thunderous smack. This behavior, called breaching, occurs often among whales. But

Whale

Gliding through the water, a whale looks as graceful as a dancer. The water supports its huge body. Despite its size, a whale can swim as if it were almost weightless. The animal can turn, twirl, and swim upside down. It uses the flippers on the sides of its body to steer. By swinging its tail fins—called flukes—up and down, the whale pushes itself through its watery world.

A whale is well equipped for life in the water. Its body is smooth and torpedo-shaped for easy movement. A thick layer of fat, called blubber, covers the animal's body and protects it from the cold temperatures of the ocean depths.

One kind of whale, the sei (say SAY) whale, has an especially sleek and streamlined body. With its pointed snout, the animal cuts through the water. Perhaps the fastest of all the whales, the sei whale can swim 23 miles (37 km) an hour in short spurts.

The champion deep diver among whales is the sperm whale. The sperm whale can dive a mile or more in search of food. It can stay underwater for longer than fifty minutes. Then, like all whales, it must surface to breathe.

The whale breathes air through a blowhole, a kind of nostril, which is located on top of its head. As a whale breaks the surface, spray shoots into the air. This spray, or spout, appears as the whale exhales. The whale's warm breath gushes out of its blowhole, hits the cooler air, and forms a cloud of mist. Different kinds of whales blow spouts in different shapes and directions.

After spouting, a whale inhales. Then by relaxing a muscle, it closes its blowhole. That way, water stays out of its nostril as the animal dives underwater again. Most great whales come up for air every 5 to 15 minutes. Since whales must surface often, they cannot sleep for very long. Instead, they doze lightly near the surface. They sometimes continue to swim as they sleep.

Scientists divide whales into two groups, toothed whales and baleen (say buh-LEAN) whales, depending on how the animals feed. Toothed whales catch their prey with peglike teeth. The largest kind of toothed whale is the sperm whale. A male grows 55 feet (17 m) long and weighs 53 tons (48 t). The sperm whale lives in all the ice-free oceans

WHALE

LENGTH OF HEAD AND BODY: 5-100 ft (152 cm-30 m)

WEIGHT: 75 lb-200 t (34 kg-181 t)

HABITAT AND RANGE: all the oceans of the world

FOOD: krill, squid, fish

LIFE SPAN: 8 to more than 80 years in the wild, depending on species

REPRODUCTION: 1 young after a pregnancy of 10 to 16 months, depending on species

ORDER: cetaceans

scientists can only guess why whales act this way. Breaching may help whales communicate or show their strength. It may help to clean their skins of pests. Whales may also leap and splash just for fun.

of the world. Like all toothed whales, it uses echolocation (say ek-oh-low-KAY-shun) to find its food.

As it swims, a toothed whale sends out pulses of high-pitched clicking sounds. When the sounds hit an object in the water—a fish or a giant squid—echoes bounce back. By listening to the echoes, a sperm whale can tell where an object is and if it is moving. In this way, it can find a meal or avoid an obstacle in its path. Porpoises use echolocation. So do bats. Read how bats use echolocation beginning on page 77. When the sperm whale locates a squid, it grabs the animal in its teeth and swallows it whole.

Scientists think that baleen whales may not use echolocation. Baleen whales such as the humpback whale and the minke (say MINK-ee) whale have no teeth. Instead, they have comblike plates called baleen in their upper jaws. The baleen serves as a huge strainer. As a whale moves through the ocean, it takes in mouthfuls of water filled with small fish or shrimplike animals called krill. The whale presses its tongue against the baleen and pushes the water out of its mouth. The food is trapped by the baleen, and the whale can swallow it.

With a flick of its tail, a humpback whale dives deep. ▷
Great whales usually stay underwater for only 5 to 15 minutes at a time. Then they must surface again to breathe. A whale breathes through a blowhole, a kind of nostril, on the top of its head. As the whale's warm breath hits the air, it turns into a moist spray, called a spout.

Gray whale: 49 ft (15 m) long

△ *Showing a mouthful of baleen, a gray whale surfaces near California. Like all baleen whales, it feeds by using the comblike bristles in its upper jaw as a strainer. Most baleen whales feed near the surface on shrimplike animals called krill. But gray whales eat tiny animals from the sea bottom.*

Baleen, sometimes called whalebone, is made of a hard, flexible material similar to that of fingernails. The size and shape of the baleen differs from one kind of whale to another. The gray whale has baleen that measures about 18 inches (46 cm) long. But the bowhead whale grows plates of baleen as long as 14 feet (4 m)!

During the summer, most baleen whales feed in the cold waters of the polar regions. There krill and other small animals often fill the sea like a soup. An adult blue whale may feast on as much as 8 tons (7,258 kg) of krill a day.

After adult baleen whales feed for four or five months—and gain as much as 40 tons (36 t)—most migrate, or travel, to warmer waters for the winter. Some gray whales of the northern Pacific swim 5,000 miles (8,047 km) from their feeding grounds near Alaska to their winter homes off the coast of Mexico.

Scientists do not know how they find their way to the same place year after year. But they follow the same path in fall and in spring. During the months away from their feeding grounds, baleen whales eat little. They feed whenever they can. But if they cannot find food, they live off their blubber.

Some toothed whales also migrate long distances when the seasons change. In their winter homes, or breeding grounds, both toothed whales and baleen whales mate and bear young. About a year after mating, a female whale gives birth to a single calf. The calf must surface immediately to breathe. Otherwise it may drown.

As a young whale nuzzles up to its mother, she squirts milk into its mouth. A calf nurses for as long as a year, growing quickly. During that time a mother and calf stay close to each other. They show affection by touching each other with their flippers.

After swimming through an underwater forest of giant kelp (below), a gray whale surfaces with a mouthful of seaweed (right). Some experts think that whales play with seaweed rather than eat it. Gray whales live in the coastal waters of the northern Pacific. They take their name from the grayish color of their skin. Many people have watched gray whales pass the coast of California as they migrate, or travel, with the change of seasons. The animals seek warmer waters in the winter and return to colder ones in the spring. Though people once hunted the gray whale extensively, laws protect the animal today.

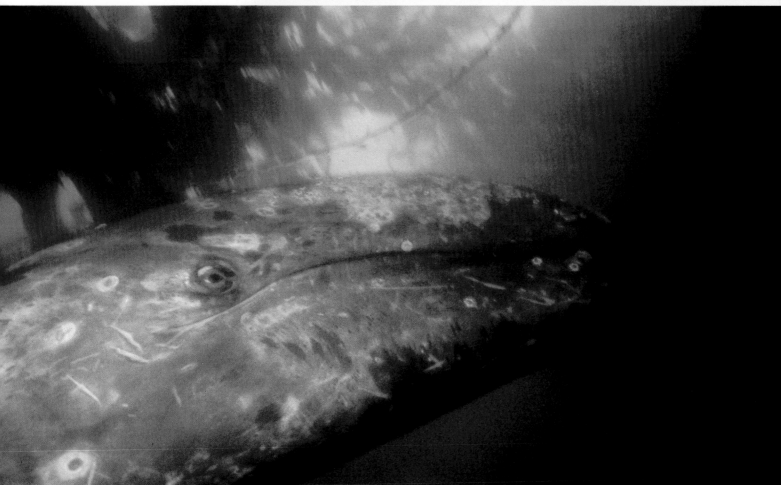

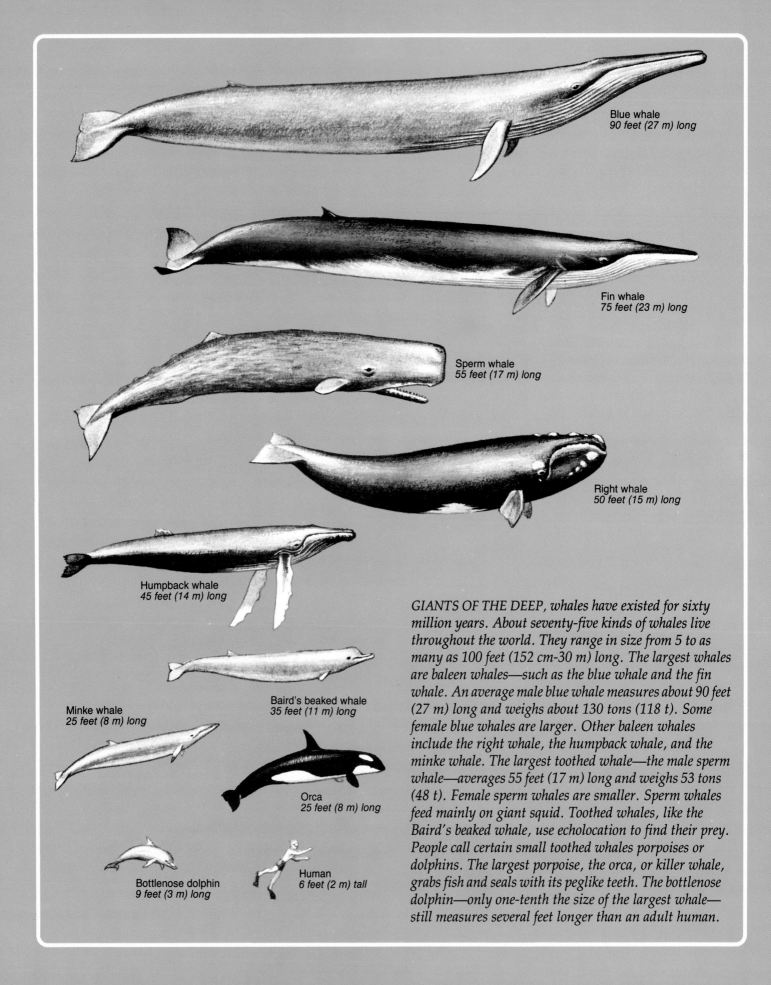

Blue whale
90 feet (27 m) long

Fin whale
75 feet (23 m) long

Sperm whale
55 feet (17 m) long

Right whale
50 feet (15 m) long

Humpback whale
45 feet (14 m) long

Baird's beaked whale
35 feet (11 m) long

Minke whale
25 feet (8 m) long

Orca
25 feet (8 m) long

Bottlenose dolphin
9 feet (3 m) long

Human
6 feet (2 m) tall

GIANTS OF THE DEEP, whales have existed for sixty million years. About seventy-five kinds of whales live throughout the world. They range in size from 5 to as many as 100 feet (152 cm-30 m) long. The largest whales are baleen whales—such as the blue whale and the fin whale. An average male blue whale measures about 90 feet (27 m) long and weighs about 130 tons (118 t). Some female blue whales are larger. Other baleen whales include the right whale, the humpback whale, and the minke whale. The largest toothed whale—the male sperm whale—averages 55 feet (17 m) long and weighs 53 tons (48 t). Female sperm whales are smaller. Sperm whales feed mainly on giant squid. Toothed whales, like the Baird's beaked whale, use echolocation to find their prey. People call certain small toothed whales porpoises or dolphins. The largest porpoise, the orca, or killer whale, grabs fish and seals with its peglike teeth. The bottlenose dolphin—only one-tenth the size of the largest whale—still measures several feet longer than an adult human.

Southern right whale: 50 ft (15 m) long when fully grown

△ *Young southern right whale dwarfs a photographer as it cruises off the coast of Australia. At about 25 feet (8 m) in length, the whale has reached only half its adult size. When fully grown, it will weigh about 40 tons (36 t). Unlike other right whales, which have mostly dark skin, this whale is a rare whitish color.*

Most kinds of whales travel together in groups, sometimes called pods. Sperm whales, though, live together in groups called harems. Several female sperm whales and their young follow one adult male. Some males swim alone, while others may travel together in pods.

Whales often communicate with one another. They make many different kinds of noises, from high-pitched squeaks to low, rumbling groans. Humpbacks sing. They repeat the same song, with slight differences from whale to whale. Over and over again, they sing their eerie combination of wails, moans, and shrieks. No one knows why.

Other kinds of whale behavior also puzzle the experts who study these huge animals. Whales often slap the water with their flippers and flukes. They leap out of the water and splash down on their backs. This activity is called breaching.

Scientists do not know why whales breach. It may be a form of communication or a way of playing. Or breaching whales may be trying to shake off annoying parasites—small animals that cling to their bodies.

Parasites can be harmful to whales. But there are bigger dangers that lurk in the sea: sharks and large porpoises called orcas. Both these animals prey on whales, particularly on the young. Orcas—also known as killer whales—hunt together in packs. Aggressive and fast, they can attack and kill an adult blue whale many times their size.

The whale's most dangerous enemies, though, are people. For centuries, human hunters have killed whales—for meat, for oil, and for baleen. Sperm whales are valued for a waxy substance called ambergris (say AM-ber-griss), located in the intestines. It is used in making some perfumes.

Until about a hundred years ago, whaling was a major industry in the United States. Whalers would set out on long voyages in pursuit of the animals. When the men on deck sighted a whale's spout, they sang out, "Thar she blows!" Then, in smaller boats, they followed the whale until they were close enough to throw their harpoons, kill the animal, and pull it in.

After several years at sea, the whaling ships

▽ *Surfacing through a hole in the ice near Antarctica, a minke whale shoots a cloud of mist into the air. The waters around it have frozen almost completely. Baleen whales such as minkes travel to polar seas in summer. They retreat to warmer waters in the winter months.*

would return to port loaded with baleen and oil melted down from blubber. The oil was made into soap, or it was used to keep lamps burning. Stiff and springy, the baleen was formed into umbrella ribs and horsewhips. It also was used in women's undergarments.

Later, more effective methods of hunting whales were developed. Whalers went to sea in large factory ships and fast boats with harpoon guns. In the animal's feeding grounds, whalers killed huge numbers of the gigantic sea mammals.

Over the centuries, whalers hunted some kinds of whales more than others. The right whale got its name because it was considered the "right" whale to hunt. It swam slowly, floated when dead, and supplied large quantities of oil. Sperm whales, gray whales, humpback whales, and other kinds were also hunted extensively. So many were killed that some species were in danger of extinction.

Today most kinds of great whales are protected. Many countries have set limits on the number of whales that can be hunted.

Minke whale: 25 ft (8 m) long

Western white-bearded wildebeest: 53 in (135 cm) tall at the shoulder

Kicking up dust, western white-bearded wildebeest bulls gallop through a woodland clearing in Tanzania.

Wildebeest

(say WILL-duh-beast)

QUIETLY GRAZING, a male wildebeest, called a bull, stands on a vast grassland in Africa. Suddenly the big, bearded animal tosses his head. He snorts, paws the ground, rolls in the dust, digs his horns in the earth, and thrashes his tail. To an observer, the wildebeest might seem to be clowning or frisking. But such a display is a way of showing that he controls a territory, or area.

The wildebeest—also known as the gnu (say NYOO)—is a large antelope. It looks like a combination of different animals. The wildebeest has a broad head and thick horns somewhat like those of a bison, a flowing tail like that of a horse, and long, thin legs like an antelope's. A mane grows on its neck and shoulders, and a coat of short hair covers its hide. The bull's deep grunts sound like a giant croaking frog. Because of its unusual appearance and spirited behavior, Dutch settlers in South Africa named the animal *wildebeest*, which means "wild beast."

A male wildebeest may weigh as much as 550

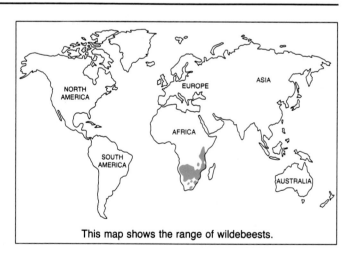

This map shows the range of wildebeests.

pounds (249 kg) and measure more than 4 feet (122 cm) tall at the shoulder. Like the male, a female, called a cow, has horns. But she is smaller.

Two species, or kinds, of wildebeests make their homes in Africa: the blue wildebeest and the black wildebeest. The grayish blue color of its coat

577

gives the blue wildebeest its name. Dark bands streak the animal's neck, shoulders, and sides. Some blue wildebeests have fringes of light-colored chin and neck hair. These animals are known as white-bearded wildebeests.

The black wildebeest has no stripes. Most of its body is dark. But its long, silky tail, sweeping almost to the ground, is a whitish color. Sometimes the animal is known as the white-tailed wildebeest.

The blue wildebeest ranges from southern to eastern Africa. The black wildebeest, which once roamed the grasslands of South Africa, nearly died out in the last century. Several thousand black wildebeests now live protected on private farms and on wildlife preserves, mostly in small herds.

Wildebeests feed mainly on grasses. They look for tender green blades. Those that live on East Africa's Serengeti Plain have no trouble finding food during the rainy season. But when the dry season

△ Still wobbly, a western white-bearded wildebeest calf rises to its feet minutes after birth. The mother cleans the newborn by licking it.

▽ With zebras nearby, western white-bearded wildebeest cows graze with their young in Tanzania. Most wildebeest calves are born during a three-week period in the rainy season.

begins, the animals there are forced to leave the parched area. They migrate, or travel, more than 800 miles (1,287 km). Always on the move, they search for green grass and water.

The migration from the plains often is heaviest in May. Then more than a million wildebeests—joined by zebras, gazelles, and other animals—wind across the Serengeti like a slow-moving train. The animals head westward and northward into open woodlands. There they can find food and water until November. As soon as the rains change the dry plains back to a green grassland, the animals return to the Serengeti. Where food and water are available all year, wildebeests usually do not migrate.

The strong bulls in every wildebeest group defend territories that they have marked with their waste. They also mark the areas with substances produced by glands on their faces and hooves. When wildebeests migrate, the bulls establish territories wherever they stop. Bulls that remain in one

Two Egyptian geese move away as a group of blue wildebeests approaches a water hole in Namibia. At certain times of the year, wildebeests often travel miles across parched plains to find food and water.

Blue wildebeest: 53 in (135 cm) tall at the shoulder

area may keep the same territories year after year. Usually, a bull must have a territory to attract females during the mating season.

When a bull claims a territory, he may be challenged by a neighboring male that trespasses. No two challenges are exactly alike. Often, the bull turns his side to the intruder and blocks his path. The two then circle each other. If the invader refuses to take part in the challenge, the first wildebeest may begin bucking, spinning, and kicking.

The two may start to graze, cautiously eyeing each other. Suddenly both drop to their knees. Thrusting their horns into the ground, they raise a

shower of dust. Sometimes they really fight. The bulls remain on their knees, lock horns, and ram each other briefly. Serious injury is rare. The intruding bull often walks away, grazing as he goes.

Young males and bulls without territories form bachelor herds. Bachelors are often forced to live on the fringes of the wildebeest group. There, among the tall grasses, enemies such as lions often lurk. If a herd of bachelors wanders into a bull's territory, the bull quickly chases the group out.

Groups of females and young roam in nursery herds. They may wander from one bull's territory to another. When a bull sees the cows approaching, he

◁ *Thousands of western white-bearded wildebeests graze on a preserve in Kenya. Even in the dry season, the area has enough water to support many animals.*

calls out with loud, deep grunts. If he fails to lure them with his calls, he may chase them and try to keep them in his territory.

About eight and a half months after mating, a female bears a single calf. The calf, able to stand shakily within minutes of its birth, can walk and run soon afterward. In a few days, it is fast enough and strong enough to keep up with the herd.

Most wildebeest calves are born during a three-week period. The calves are not strong enough to outrun such enemies as spotted hyenas, lions, cheetahs, leopards, and wild dogs. But the presence of many calves helps protect the young during their first few days. Their enemies may have trouble singling out one calf to prey on.

Find out about some relatives of wildebeests—gazelles, gerenuks, and impalas—in their own entries. Read about other antelopes on page 52.

WILDEBEEST

HEIGHT: **45-55 in (114-140 cm) at the shoulder**

WEIGHT: **350-550 lb (159-249 kg)**

HABITAT AND RANGE: **grassy plains and open woodlands in southern, central, and eastern Africa**

FOOD: **mainly grasses**

LIFE SPAN: **as long as 21 years in captivity**

REPRODUCTION: **1 young after a pregnancy of about 8½ months**

ORDER: **artiodactyls**

Broomlike tufts of hair stand straight up on the neck and face of a black wildebeest. This animal's short, stiff mane, handlebar horns, and light-colored tail set it apart from its closest relative, the blue wildebeest. Once hunted nearly to extinction, black wildebeests now live protected on preserves and private farms.

Black wildebeest: 47 in (119 cm) tall at the shoulder

581

Wolf

HEAD THROWN BACK, a wolf points its nose toward the sky and begins to howl. The call soars to a high note, then slides down in smooth, rippling tones. Wolves may howl at any time of the day or night.

Wolves keep in touch by howling. A rising and falling "lonesome howl" may mean that an animal has become separated from the pack, or group. A howling pack may be warning other packs to stay away from its hunting grounds. Wolves also howl to call the pack together after a hunt is over. It often seems that wolves howl simply for the pleasure of being together. When one wolf starts to howl, other members of the pack join in.

The wolf is the largest member of the dog family. It looks much like a German Shepherd dog. It has thick, shaggy fur and a bushy tail. With its long legs, it can run great distances. And a wolf's powerful jaws can seize prey and hold it tightly.

Wolves can have coats of different colors. The gray wolf usually has a coat that is a mixture of white, black, gray, and brown hairs. Sometimes it may have black fur. Other gray wolves may have reddish coats. Gray wolves that live on the tundra, the treeless plains of the Far North, have thicker, longer fur that may be almost white. These wolves are called tundra wolves.

Another kind of wolf, the red wolf, has shorter fur than the gray wolf. Some red wolves do have reddish coats, but others vary in color from light tan to gray or black.

Wolves once roamed almost the entire world north of the Equator. They were at home everywhere, except in tropical regions and in deserts. They lived in forests and on prairies, grasslands, and tundra.

Today gray wolves—often called timber wolves—have been hunted almost to extinction in the United States, except in Alaska and Minnesota. A few packs also survive in Michigan and Wisconsin. Wolves are rare elsewhere in the world, except

◁ *Casting long shadows on the snow, gray wolves cross a frozen lake in Minnesota. The pack follows a trail made on earlier rounds of hunting deer and moose. In deep snow, wolves usually travel in single file. Each wolf can follow the trail more easily, because it does not have to break its own path through the drifts.*

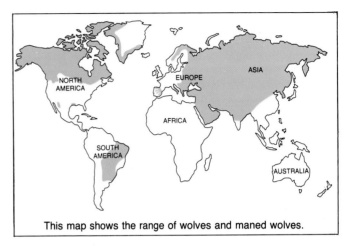

This map shows the range of wolves and maned wolves.

WOLF AND MANED WOLF

LENGTH OF HEAD AND BODY: 36-63 in (91-160 cm); tail, 13-20 in (33-51 cm)

WEIGHT: 40-175 lb (18-79 kg)

HABITAT AND RANGE: forests, tundra, swamps, and prairies in parts of North America, Europe, and Asia. Maned wolf: grasslands and swamps in South America

FOOD: large and small mammals, birds, fish, lizards, snakes, frogs, insects, and fruit

LIFE SPAN: 16 years in captivity

REPRODUCTION: 5 to 7 young after a pregnancy of about 2 months

ORDER: carnivores

Whining wolves wait as their pack leader eats. Wolves may go for days without making a kill. When they bring down large prey, they gorge themselves. A wolf may eat almost 20 pounds (9 kg) of meat in one meal.

Gray wolf: 52 in (132 cm) long; tail, 16 in (41 cm)

in Canada and many parts of Asia. Some are scattered in Mexico and Europe. A few red wolves may remain in wet grasslands in parts of Texas and Louisiana. Distant relatives called maned wolves live on grasslands and in swamps in South America.

Few animals have been feared so much by people as the wolf. But there is little reason for such fear. Unless a wolf is sick, it will not usually attack a human. Yet, for centuries, people have hunted, poisoned, and trapped wolves.

The biggest threat to wolves is their changing habitat. As settlers move into the wilderness areas where wolves live, people may raise livestock there. Wolves sometimes attack the livestock. Then farmers and ranchers try to get rid of the wolves.

Wolves will eat almost anything—small mammals, birds, fish, lizards, snakes, and fruit. They usually depend on large prey, however—moose, elk, caribou, sheep, and deer—for their food.

Because their prey is often so much larger than they are, wolves must hunt together in packs. Tails wagging, the wolves crowd around the pack leader. They frisk about, touch noses, and lick the leader's mouth. A few whines and a long, low howl may signal the beginning of the hunt. The members of the pack all join in the call. Each animal howls on a different note. The excitement peaks in a wild chorus.

The wolves set out at a smooth, easy trot. Wolves can lope along for many miles, and they can run quickly for short distances. Usually wolves

On a hilltop in a national park in Italy, a pack of gray wolves rests and stretches during a winter afternoon.

△ Nose pointed skyward, a gray wolf fills the frosty air with a howl. A wolf can vary its call by sucking in its cheeks. It can also change the sound by curling and uncurling its tongue.

Nose to nose, two gray wolves gently nuzzle each ▷ other. The wolf with the lighter fur is the female leader of a pack in Minnesota.

travel in single file. In winter, they may travel great distances to find enough food.

Wolves try to attack any prey they find. Strong and healthy animals can often escape, and many chases end in failure. The victims caught are usually sick, injured, or very young or old animals.

Wolves hunt different animals in different ways. They may chase a caribou herd until they spot a weak animal and attack it. Mountain sheep may try to escape from wolves by running up steep, rocky cliffs. Wolves can surprise them by attacking from above. When musk-oxen are threatened, they form a line to protect themselves. Generally, the only way wolves can bring down a musk-ox is to find an animal that has been left alone.

A moose is a wolf's largest and most dangerous prey. A moose can be ten times as large as a wolf. Charging and whirling, a moose kicks at attackers with its hind feet and slashes with its front feet. But if the wolves can make a moose run, the larger

animal cannot easily kick. Then the wolves can bring it down by attacking from behind.

A wolf pack often includes about six wolves. All the members are usually related. One male acts as the leader of the entire pack. One female leads the females and the young. In larger packs, she also leads less important males. Each wolf has a place in the pack. A leader shows its rank by holding its head up and raising its tail straight up. Less important wolves roll over, wriggle, or crouch before the leader. They lay back their ears and tuck their tails between their legs. Sometimes pack members bare their teeth, growl, or snap at one another. But these threats seldom turn into real attacks.

The two leaders keep the pack together and lead the defense against such enemies as bears or wolves in other packs. The leaders decide when to hunt, and they choose the prey. They also settle fights over food or between cubs. The leader is usually the first to feed after a kill. The male and female leaders are often the only members of the pack to mate and bear young.

Before the young are born in the spring, the male and the female prepare a den. Sometimes

Brightly colored sockeye salmon attract a gray wolf in ▷ Alaska. The water churns with fish swimming upstream to lay eggs. Wolves catch fish in shallow streams by snapping them up in their jaws. Despite the plentiful prey, this wolf needed several bites to land a meal.
▽ Red wolf follows a trail through the brush. Now in danger of extinction, red wolves once roamed parts of the southeastern United States.

Red wolf: 47 in (119 cm) long; tail, 15 in (38 cm)

wolves dig a new den. They may also enlarge a fox den or use a beaver lodge. A wolf den may be 15 feet (5 m) long and high enough for a wolf to stand in.

Usually from five to seven pups are born in a litter. A newborn wolf has short legs and a blunt nose. The offspring is covered with dark fur. When the pups are about two weeks old, their eyes open. At about three weeks, they crawl out of the den for the first time. Then they begin to eat solid food. They

may continue to nurse for another month, however.

The entire pack takes an interest in a new litter. All the adults help care for the pups, bringing food to them and sometimes baby-sitting when their parents are hunting. When a wolf returns from hunting, the pups rush up to it. They whine and wag their tails and lick the adult's face. The wolf then brings up some of the meat it has swallowed and brought back to the den in its stomach.

Soon the pups start exploring the world around their den. Rough-and-tumble play helps decide which young wolves will become leaders of the pups. One pup sneaks up and pounces on another. That sets off a wild chase. Pups attack each other in play. They practice hunting by attacking insects, birds, and rodents. These games help the pups learn hunting skills they will need later on.

During the summer, the pups stay in several

Maned wolf: 42 in (107 cm) long; tail, 15 in (38 cm)

resting spots while the adults hunt. By fall, the young wolves may be able to join the pack as it travels. They help to run down prey, but they usually let the older wolves make the kill.

When winter arrives, the young are nearly grown. But they will not be expert hunters until they are almost two years old. Then some young wolves will stay with the pack. Others will leave to find mates and to start new packs.

◁ *Maned wolf, another member of the dog family, gallops through tall grass in South America. A black mane stands up on its neck, and black hairs cover its muzzle and legs. Maned wolves hunt small mammals and birds.*

Wolverine

(*say* WOOL-vuh-reen)

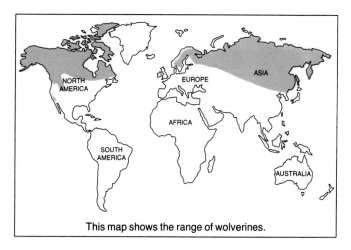

This map shows the range of wolverines.

WOLVERINE

LENGTH OF HEAD AND BODY: 26-34 in (66-86 cm); tail, 7-10 in (18-25 cm)

WEIGHT: 24-40 lb (11-18 kg)

HABITAT AND RANGE: forests and tundra in northern North America, Europe, and Asia

FOOD: remains of dead animals, large hoofed animals, small mammals, birds, eggs, fruit, and plants

LIFE SPAN: as long as 16 years in captivity

REPRODUCTION: 1 to 4 young after a pregnancy of about 6 months

ORDER: carnivores

AN AIR OF MYSTERY has long surrounded the wolverine. People rarely see this furry, bearlike animal with the short, bushy tail. It is legendary for its fierceness, its great strength, and its huge appetite. The wolverine lives only in wilderness areas—forests and treeless plains—in parts of northern North America, Europe, and Asia.

Despite its appearance and the sound of its name, the wolverine is not related to the bear or to the wolf. One of the largest members of the weasel family, the wolverine may weigh 40 pounds (18 kg). It measures 3 feet (91 cm) long, including its tail.

In summer, the wolverine eats a wide variety of food—berries, plants, eggs, rodents, and rabbits. In winter, the wolverine feeds mainly on dead animals it finds, such as caribou, elk, and deer. With its keen sense of smell, the wolverine can find food buried under snow. It may also dig some rodents out of the burrows where they hibernate (say HYE-bur-nate), or sleep, all winter. The wolverine may kill more than it can eat at one time. It stores the meat and returns to it later.

Occasionally, a wolverine may attack a caribou that is weak or bogged down by snow. With its furry paws, the wolverine can move quickly across snow, when travel is difficult for larger animals.

As it looks for food, the wolverine lopes slowly

△ On a midsummer afternoon, a wolverine hunts for rodents along a ridge. The animal, a large member of the weasel family, is legendary for its great strength and fierceness.

△ Blanketed by heavy, dark fur, a male wolverine in Alaska pauses in the snow.

throughout a wide range. It often travels as far as 15 miles (24 km) in a day. The wolverine marks its range with a strong-smelling substance produced by glands in its body. The scent marks may mean, "I am hunting here." The animal also marks trees with scratches and bites.

A male shares his range with two or three females. In late winter or early spring, about six months after mating, the female makes a den under the snow or in a hidden place. There she gives birth to a litter of one to four young. They are fully grown in about six months, and by the following winter they go off on their own.

Trappers in North America once hunted wolverines for their fur, which was used to line parkas. By the early 1900s, the animals had almost disappeared. Since the 1960s, wolverines have been protected by law in several states.

Wombat

(say WAHM-bat)

Hairy-nosed wombat: 36 in (91 cm) long; tail, 2 in (5 cm)

Hairy-nosed wombat waddles among clumps of grass in its feeding grounds. The plump marsupial's front teeth have sharp edges that can easily cut tough grasses, roots, and bark.

ROLY-POLY AS A BEAR CUB, a stocky wombat basks in the early morning sunlight. Then this marsupial (say mar-soo-pea-ul), or pouched mammal, gets up and waddles into its nearby burrow. It curls up there and sleeps most of the day. At night, the wombat heads for its feeding grounds, where it searches for roots, grasses, and bark to eat.

Wombats live in forests and on grasslands in parts of Australia and its nearby islands. There are two kinds of wombats. The coarse-haired wombat, which weighs as much as 80 pounds (36 kg), has rough brown or black fur. The smaller hairy-nosed wombat has a silky, gray-and-brown coat.

Like badgers, wombats dig large burrows with their short, powerful legs and sharp claws. They line their dens with pieces of bark.

When alarmed, the shy wombat runs into its burrow. But if a dingo, a kind of wild dog, tries to

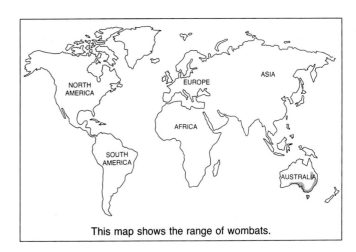

This map shows the range of wombats.

follow it inside, the wombat kicks the attacker with its strong hind legs.

A female wombat gives birth to one tiny off-spring between April and June of each year. The

590

young wombat crawls into the rear-opening pouch on its mother's belly and continues to grow and develop there. As the mother walks about, her offspring can sometimes be seen peeking out between her hind legs.

At about five months of age, the young wombat begins to leave the pouch for short periods. It returns frequently to nurse or to seek shelter. By December, it is able to take care of itself.

WOMBAT

LENGTH OF HEAD AND BODY: 28-47 in (71-119 cm); tail, as long as 2 in (5 cm)

WEIGHT: 32-80 lb (15-36 kg)

HABITAT AND RANGE: forests, grasslands, and dry regions in parts of Australia and nearby islands

FOOD: grasses, roots, and bark

LIFE SPAN: about 25 years in captivity

REPRODUCTION: 1 young after a pregnancy of about 1 month

ORDER: marsupials

Deep burrows on a dry plain in Australia provide homes for a colony of hairy-nosed wombats.

Woodchuck

The woodchuck is a kind of marmot. Find out about marmots on page 358.

Yak

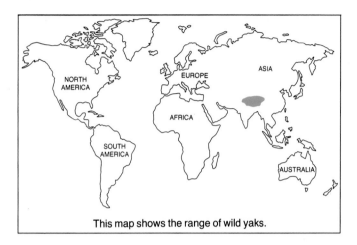
This map shows the range of wild yaks.

THE WINDSWEPT HIGHLANDS OF TIBET provide a home for the wild yak of central Asia. Few other large mammals live at such heights. The massive yak roams cold, treeless plateaus and the mountain ranges nearby. It can climb as high as 20,000 feet (6,096 m) above sea level.

The yak has a low-slung body, and it often keeps its broad head close to the ground. Males have long, curving horns that may measure 3 feet (91 cm) from tip to tip. When fully grown, male wild yaks may reach 6 feet (183 cm) at the shoulder and weigh as much as 1,800 pounds (816 kg). Females are much smaller and have shorter, thinner horns.

Surefooted relative of the cow, the sturdy yak is suited to the harsh lands where it lives. During the winter, a soft, dense undercoat grows beneath coarse, longer hairs and helps protect the animal from the cold. In spring, yaks shed their woolly undercoats. Often, as the wool falls out, chunks of it get caught in the long fringes that hang from their shoulders, sides, and legs.

Yaks graze on grasses and herbs. They also browse, nibbling the leaves of small shrubs. Like cows, they do not chew their food thoroughly before swallowing. After eating, they bring up wads of partly digested food, called cuds. The animals chew the cuds further, then swallow and digest them.

Female yaks and their young travel in herds

▽ Sturdy and surefooted, shaggy domestic yaks travel a narrow trail in Nepal. They climb near Mount Everest, at 29,028 feet (8,848 m) the world's highest mountain.

Casting shadows across the snow, yaks and porters approach a mountain pass in Pakistan.

that may number hundreds of animals. In such large groups, they can protect themselves from wolves. Yaks have a keen sense of smell. If members of the herd pick up a scent of danger, they may rush together. They face the enemy with heads lowered and defend themselves with their horns.

Older males usually stay with the females only during the mating season, in the fall. During the rest of the year, males roam alone or in small groups.

About nine months after mating, a female bears one calf. It remains with her for at least a year.

During the hottest part of the summer, yaks leave lower pastures, where they spend the winter. They head for cool mountain heights, where snow remains on the ground all year. Yaks are skilled at feeding in the snow. They brush it aside with their muzzles or use their hooves to uncover patches of grass. When water is scarce, they may eat the snow.

Centuries ago, people of Asia began to domesticate, or tame, the yak. Domestic yaks are much smaller than wild yaks. Their black or brown coats may have patches of white or red. Domestic yaks make sturdy pack animals. They supply their owners with milk, meat, hair, and hides for leather. Their waste is burned for fuel.

YAK

HEIGHT: 37 in-6 ft (94-183 cm) at the shoulder

WEIGHT: 400-1,800 lb (181-816 kg)

HABITAT AND RANGE: mountainous regions of central Asia; domestic yaks are found throughout a wider area of central and southern Asia

FOOD: grasses, herbs, leaves, shoots, and twigs

LIFE SPAN: about 20 years in captivity

REPRODUCTION: 1 young after a pregnancy of about 9 months

ORDER: artiodactyls

Z

Zebra

(say ZEE-bruh)

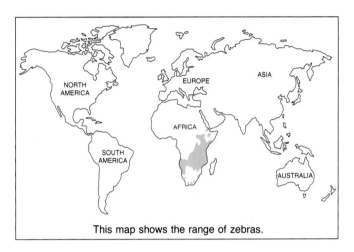

This map shows the range of zebras.

NO TWO ZEBRAS have stripes that are exactly alike. Each set of stripes is a little different from the next, just as one person's fingerprints are always different from another's. Even the patterns on opposite sides of a zebra's body do not match exactly.

No one knows why zebras have stripes. They may help hide the horselike animals in tall grass or in shimmering sunlight. When a group of zebras runs, it may be hard for an enemy to pick out a single animal. The different patterns may also help zebras tell each other apart. Scientists think the stripes may help protect the animals against some biting flies. These insects can only make out large patches of black or white. They probably do not see the zebras' narrow stripes, so they rarely bother the animals.

People have always been fascinated by the zebra's stripes. Centuries ago, Romans paraded captured zebras through the streets of Rome. Their name for the zebra meant "tiger horse."

Today three kinds of zebras live in Africa. The most numerous are plains zebras, which are found in open, grassy areas. Plains zebras sometimes are called Burchell's zebras. Mountain zebras live on rocky hillsides. People rarely see these animals. Grevy's (say GRAY-veez) zebras are larger than the

Tail waving, a plains zebra in eastern Africa leaps into the water as it crosses a stream.

594   Plains zebra: 51 in (130 cm) tall at the shoulder

△ *Nipping and nuzzling, young plains zebras play at fighting. Zebras often groom each other. They sometimes rest with their heads across each other's backs.*
▽ *Although no two zebras' stripes match exactly, each kind of zebra has a similar pattern. Short stripes run crosswise over the rump of a mountain zebra. Narrow, black stripes, spaced close together, mark a Grevy's zebra. A plains zebra has broader stripes. Fainter bands sometimes alternate with the darker stripes.*

Plains zebra

Grevy's zebra

Mountain zebra

△ *Flying hooves send up clouds of dust as plains zebras move across a dry grassland in Africa.*
Antelopes called springboks travel with the group. If alarmed, zebras bunch together for protection.

others. They live on dry plains and on hillsides.

Plains and mountain zebras—no bigger than ponies—roam in family groups. A male, called a stallion, stays with as many as eight females, called mares, and their young. Even when families graze in herds of as many as 10,000 animals, each family stays together. If family members become separated, they call to one another with barking cries.

Each stallion protects the females and young in his family. Ears pricked forward, he listens for enemies—lions, hyenas, and wild dogs. If danger threatens, the mares bunch together and hurry away with their offspring. The stallion follows, fighting off the attacker with bites and kicks.

A stallion also stays alert for rival stallions. If two males meet, they greet each other by sniffing.

They touch noses and leap apart. If one tries to steal a mare from the other to start a family of his own, the two may kick, bite, and wrestle neck to neck.

At night, plains zebras remain on flat, grassy plains. There they can see in every direction over the

ZEBRA

HEIGHT: 41-59 in (104-150 cm) at the shoulder

WEIGHT: 440-880 lb (200-399 kg)

HABITAT AND RANGE: mountains and open, grassy plains in eastern, central, and southern Africa

FOOD: grasses

LIFE SPAN: about 25 years in captivity

REPRODUCTION: usually 1 young after a pregnancy of 11½ to 13 months

ORDER: perissodactyls

△ *Plains zebras and antelopes called gemsboks gather at a water hole in Namibia. One zebra kneels to lick salt from the ground. Zebras often share water holes with other animals.*

▽ *Plains zebra mare rolls in the dust to rid herself of insects. Her days-old offspring stands close by. Young zebras can run and play just an hour after birth.*

short grass. At least one adult acts as a guard and stays awake. Just after sunrise, the zebras move off in single file to graze. They feed on tall grasses too tough for other animals. They drink at water holes that dot the plains. Elephants, wildebeests, gazelles, buffaloes, and giraffes may feed and drink nearby. With so many eyes, ears, and noses alert for danger, all the animals at a water hole are safer.

Grevy's zebras live in different kinds of groups than other zebras. Although they may come together to graze, they do not always stay together. Some individuals even travel on their own.

During certain months of the year, little rain falls in parts of Africa. Then the herds of zebras wander widely looking for food. Pawing the ground, mountain zebras may dig for water. When there is plenty of grass to eat, zebras usually remain in one place. Most young are born then. A newborn zebra's stripes are brown and white. Its mane is still wispy. When it is an hour old, the baby can frisk and run. For the first few days, its mother keeps other members of the group away until the foal recognizes her.

Zebra

Narrow stripes of Grevy's zebras blend ▷ with tall grass in the noonday sun. These zebras have long, hairy ears, and they bray like donkeys. Though other kinds of zebras live in the same family group all year round, Grevy's zebras move freely from group to group.

Grevy's zebra: 59 in (150 cm) tall at the shoulder

Zebu

The zebu is a kind of cow. Read about cows on page 158.

Zorilla

(say zuh-RILL-uh*)*

In a rare daytime appearance, a zorilla in Africa hunts for insects or rodents. The zorilla roots for prey with its snout or digs for food with its long claws.

ZORILLA

LENGTH OF HEAD AND BODY: 13-15 in (33-38 cm); tail, 8-12 in (20-30 cm)

WEIGHT: 2-3 lb (1 kg)

HABITAT AND RANGE: brushy plains and grasslands in parts of Africa

FOOD: insects, small mammals, reptiles, birds, and frogs

LIFE SPAN: 5 years in captivity

REPRODUCTION: 2 or 3 young after a pregnancy of 5 weeks

ORDER: carnivores

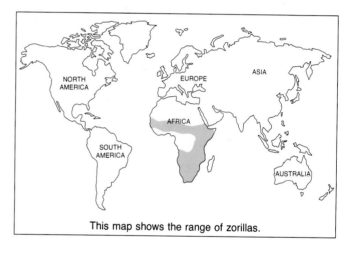

This map shows the range of zorillas.

WITH A THICK, BLACK COAT marked by white stripes, the zorilla of Africa resembles its relative the skunk. Like the skunk, the zorilla has scent glands under its bushy tail. When alarmed, the animal raises the hair on its body and tail, making itself look bigger. If that doesn't frighten its attacker away, the zorilla screams, lifts its tail, and sprays a strong-smelling fluid at its enemy.

At night, the zorilla trots across open grasslands and brushy plains on its short legs. It searches for insects, small mammals, snakes, and birds to eat. During the day, the zorilla rests in a rock shelter or in a shallow burrow.

A female gives birth in the burrow or in another well-hidden place. Usually two or three young are born after a pregnancy of five weeks.

Glossary

The *Book of Mammals* may contain words that are new to you. Their meanings are explained below. Knowing what these words mean may help you to better understand the *Book of Mammals* and other books about animals.

adapt—to become suited to surroundings. Animals adapt to changes in environment over many generations.

aestivate—to spend the hot months in a sleeplike state. While aestivating, an animal breathes more slowly and has a lower body temperature and slower heart rate than while active.

amphibians—cold-blooded vertebrates, such as frogs, toads, and salamanders. Amphibians spend part of their lives in the water and part on land.

antlers—a pair of solid bony forms on the heads of most species of male deer. Antlers sprout, continue to grow, and fall off every year. Female caribou also grow antlers.

bachelor herd—a group formed by young male animals and males without territories. Bachelor herds are common among many kinds of hoofed mammals.

baleen—plates of hard, flexible material, with fringed inner edges, that grow from the upper jaws of some whales. Baleen plates serve as strainers. They allow water to pass through a whale's mouth, but they trap small animals for food.

brachiate—to swing with the arms in a hand-over-hand movement from one hold—often a branch or a vine—to another.

breed—to mate and produce young. Also, a breed is a group of domestic animals of one species in which certain characteristics, such as size, color, and type of hair, have been developed and maintained by humans.

browse—to feed on the leaves, twigs, and young shoots of shrubs and trees.

brush—a growth of small trees or shrubs.

burrow—a hole in the ground often dug by an animal and used as a place to live, to hide, and to bear young.

bush—an area covered with low shrubby plants.

camouflage—a disguise, such as body coloring, that helps an animal blend in with its surroundings.

captivity—the state of being confined, as when an animal is kept in a zoo or an animal park.

cud—a wad of partly chewed and digested food brought up to the mouth from the stomach of certain hoofed mammals. The cud is chewed again, swallowed, and then further digested.

den—a hollow space that an animal uses as a place to live, to hide, and to bear young.

desert—a dry, generally treeless region that usually receives less than 10 inches (25 cm) of precipitation a year. A desert may be hot or cold.

displays—actions that serve as signals to other animals. Animals may use displays to attract mates. Displays may also help to ward off enemies and to show how important or strong an animal is to the other animals in its group.

domestic animals—animals, such as dogs, cats, cattle, or sheep, that have been tamed by humans.

echolocation—a system by which an animal sends out beeps or pulses of high-pitched sounds and listens for the echoes that bounce back when the sounds hit an object. Bats, porpoises, solenodons, whales, and some shrews use echolocation to find their way, to avoid obstacles, and to detect prey.

endangered—reduced in numbers to the point of near extinction.

environment—surroundings, including air, land, water, and living things.

extinct—no longer in existence.

feral animal—an animal, now living in the wild, that was once domestic or had ancestors that were once domestic.

flipper—a broad, flat limb used for swimming. Seals, porpoises, and whales have flippers.

flukes—tail fins or the flattened parts of a whale's tail.

food chain—a sequence of plants and animals that provide food for members of a community. Plants make their own food and are eaten by plant-eating animals. These animals are consumed by meat-eating animals. For example, grass may be eaten by a mouse. The mouse may be eaten by a weasel. The weasel may be eaten by a hawk.

forest—an area with thickly growing trees and underbrush.

freeze—to become very still.

grassland—an area covered with grasses and herbs.

graze—to feed on grasses and herbs.

groom—to clean the skin and hair by removing dirt and insects. For many animals, grooming one another is important in maintaining the social relations of their group.

guard hairs—long, coarse hairs that cover and protect an animal's soft, thick underfur.

habitat—the terrain in which an animal lives naturally, such as a desert, a forest, a grassland, or a swamp.

herd—a group of wild or domestic animals of one or more kinds that feed and travel together.

hibernate—to spend the cold months in a sleeplike state. While hibernating, an animal breathes more slowly and has a lower body temperature and slower heart rate than while active.

home range—the entire area where an animal lives.

hoof—a hard fingernail-like covering on the lower part of the foot of certain mammals, such as horses and deer.

horn—a hard growth made of a hairlike material that forms around bony cores on the heads of many hoofed mammals. Horns are permanent. Unlike antlers, they do not fall off and grow back every year.

incubation period—the time during which an animal uses the warmth of its body to hatch eggs. Female echidnas and platypuses are the only mammals that lay eggs.

invertebrates—animals without backbones, such as jellyfishes, worms, snails, spiders, and insects.

krill—small, shrimplike animals found most abundantly in cold waters. Krill are the main food of baleen whales.

larvae—wormlike forms that hatch from the eggs of many kinds of insects.

life span—the average or recorded length of time that an animal lives in the wild or in captivity.

litter—the offspring born at one time to an animal that usually gives birth to more than one young.

mammal—a warm-blooded vertebrate that feeds its young with milk from special glands in the mother's body. All mammals except monotremes give birth to living young. All mammals have some hair on their bodies.

mammalogist—a scientist who studies mammals.

marsh—an area of usually wet land covered with plants, such as grasses, cattails, and rushes.

mating season—the time of year when males and females of the same species breed.

migration—the seasonal movement of animals from one place to another.

molt—to shed hair at certain times of the year.

nurse—to feed young with milk from special glands in the mother's body.

nursery herd—a group formed by female animals and their young. Nursery herds are common among many kinds of hoofed mammals.

offspring—the young of an animal.

order—a scientific classification that groups together animals that share certain characteristics. An order is divided into families, genera, and species.

plain—a large area of treeless land that is either flat or rolling.

plateau—a large, flat, elevated area of land. A plateau is higher than the surrounding land on at least one side.

poacher—a person who hunts illegally.

prairie—a large area of flat or rolling land with tall, coarse grasses and few trees.

predator—an animal that hunts, kills, and feeds on other animals. Wild cats, for example, are predators of rodents.

prehensile tail—a tail that can be used for grasping. Some kinds of monkeys have prehensile tails that can hold on to branches and support their weight. Binturongs, kinkajous, tree pangolins, and most opossums also have prehensile tails.

preserve—a natural area set aside for the protection of animals.

prey—an animal that is hunted for food.

rain forest—a tropical area having a heavy annual rainfall. Rain forests have broad-leaved evergreen trees.

reproduction—the process by which animals produce offspring like themselves.

reptiles—air-breathing, cold-blooded vertebrates—such as snakes, turtles, and lizards—that are usually covered with scales or bony plates.

root—to dig in the ground with the snout.

savanna—a grassland in tropical regions that contains scattered trees or bushes.

scavenger—an animal that feeds on the remains of dead animals that it finds.

scent glands—glands in the bodies of some animals that produce a substance that is often strong smelling.

scent marks or scent posts—marks made by an animal rubbing its scent glands against an object. Scent marks or scent posts may also be made by spraying urine. Animals often use marks to announce their presence and to help find mates.

scrubland—an area covered with shrubs or stunted trees.

species—a group of animals of the same kind that can mate and produce young like themselves.

sprint—to run with a burst of speed for a short distance.

swamp—an area of wet land covered with trees and shrubs.

terrain—the physical appearance of an area of land.

territory—an area that an animal, or a group of animals, lives in and defends from others of the same species.

tundra—a relatively flat, treeless plain found in arctic and subarctic areas. Cold desert is another term for tundra.

tusks—the large teeth of an animal that usually stick out when its mouth is closed. Elephants, walruses, wild boars, and male musk deer have tusks.

underbrush—bushes, small trees, or shrubs that are found growing among the tall trees of a forest.

vertebrates—animals with backbones, such as mammals, birds, fishes, amphibians, and reptiles.

water hole—a pool where animals gather to drink.

wingspan—the distance from one wing tip to the other, when both wings are extended.

woodland—a grassland with small trees and shrubs.

Acknowledgments

The Special Publications Division is grateful to the individuals and organizations listed here for their generous cooperation and assistance during the preparation of the two-volume *Book of Mammals:* David Horr Agee, J. Scott Altenbach, American Museum of Natural History, American Rabbit Breeders Association, Inc., Paul K. Anderson, Sydney Anderson, Renate Angermann, Kenneth B. Armitage, Rollin H. Baker, Julie Ball, Marion Ball, Carl Brandon, Anton B. Bubenik, John A. Byers, Carnegie Museum of Natural History, James W. Carpenter, Tracy S. Carter, Joseph A. Chapman, Tim W. Clark, Garrett C. Clough, Malcolm Coe, C. G. Coetzee, Richard W. Coles, Larry R. Collins, Ian McTaggart Cowan, Dorcas D. Crary, A. W. Crompton, Anne Dagg, Robert J. Davey, Joseph A. Davis, Jerome E. DeBruin, James G. Doherty, James M. Dolan, Jr., Iain Douglas-Hamilton, Duke University Center for the Study of Primate Biology and History, Nicole Duplaix, Harold J. Egoscue, John F. Eisenberg, Elephant Interest Group/Department of Biological Sciences/Wayne State University, Margaret Ellis, Robert K. Enders, Albert W. Erickson, Purcell Erquhart, Richard D. Estes, M. Brock Fenton, Fish and Wildlife Service of the U. S. Department of the Interior, Florida State Museum, Dian Fossey, Bristol Foster, Richard R. Fox, Hans Frädrich, George W. Frame, Lory Frame, William L. Franklin, Biruté M. F. Galdikas, Alfred L. Gardner, Valerius Geist, Hugh H. Genoways, Patricia Goodnight, Edwin E. Goodwin, David R. Gray, Mervyn Griffiths, Barbara Grigg, Heinz Heck, Steven R. Hill, Hendrik N. Hoeck, Judith Hopper, Maurice G. Hornocker, Nicholas Hotton III, Sandra L. Husar, Mary Alice Jackson, Jennifer U. M. Jarvis, A. J. T. Johnsingh, Wiletta V. Jones, Charles J. Jonkel, Beatrice Judge, Lloyd Keith, Karl W. Kenyon, Charles A. Kiddy, John Kirsch, Devra Kleiman, Hans Klingel, Karl Koopman, Donald L. Kramer, Karl R. Kranz, Andrew Laurie, Sam Lewis, Boo Liat Lim, Raymond Linder, Muriel C. Logan, Charles A. Long, Nicholas J. Long, William Lopez-Forment, Dale F. Lott, Yael Lubin, Richard E. McCabe, Audrey McConnell, Michael E. McManus, Mark MacNamara, Phyllis R. Marcuccio, Joe T. Marshall, James G. Mead, L. David Mech, J.A.J. Meester, Derek A. Melton, Arlene Michelirer, Russell A. Mittermeier, Patricia D. Moehlman, Peter L. Munroe, James Murtaugh, Norman Myers, National Museums of Canada, National Zoological Park/Smithsonian Institution, A. E. Newsome, New York Zoological Society, Kenneth S. Norris, Alvin Novick, Ronald M. Nowak, Margaret A. O'Connell, Bart W. O'Gara, Robert T. Orr, Ed Peifer, Jr., Michael Pelton, R. L. Peterson, Ivo Poglayen-Neuwall, Roger A. Powell, George Rabb, Mary Rabb, Urs H. Rahm, Adele S. Rammelmeyer, Galen B. Rathbun, G. Carleton Ray, Elizabeth H. Rinker, Miles Roberts, Jonathan Rood, David T. Rowe-Rowe, Daniel I. Rubenstein, James K. Russell, Hope Ryden, Oliver A. Ryder, George B. Schaller, Victor B. Scheffer, Jan Schwartz, John Seidensticker, Geoffrey B. Sharman, James H. Shaw, Paul W. Sherman,

On an ice floe in Canada's Gulf of St. Lawrence, a female harp seal stretches out in the late-winter sun with her white-coated, week-old pup. The pup will nurse for less than two weeks. It will gain weight rapidly and soon weigh about 100 pounds (45 kg). By spring, the pup will be able to care for itself.

Jeheskel (Hezy) Shoshani, A.R.E. Sinclair, R.H.N. Smithers, the Smithsonian Institution, Smithsonian Tropical Research Institute, Lyle K. Sowls, R. E. Stebbings, Eleanor E. Storrs, Michael D. Stuart, Fiona Sunquist, Mel Sunquist, Lee M. Talbot, Ian Tattersall, John J. Teal, Geza Teleki, Clair E. Terrill, Richard W. Thorington, Jr., Merlin D. Tuttle, University of Maryland's Appalachian Environmental Laboratory, U. S. Department of Agriculture/Beltsville Agricultural Research Center, Richard G. Van Gelder, Franklyn Van Houten, C. G. van Zyll de Jong, Charles Walcott, John S. Ward, Janet Warner, Everett J. Warwick, Washington University Tyson Research Center, Christen M. Wemmer, Ralph M. Wetzel, Robert J. Whelan, Judith White, Wildlife Management Institute, David P. Willoughby, Henk Wolda, Charles A. Woods, the World Armadillo Breeding and Racing Association, World Wildlife Fund, William A. Xanten, Hoi Sen Yong.

Additional Reading

You may want to check the National Geographic Society Index in a school or a public library for articles about mammals and to refer to the books below. ("J" indicates a juvenile book.)

GENERAL:
Buchenholz, Bruce, *Doctor in the Zoo*, 1974 (J).
Burt, William H., *A Field Guide to the Mammals*, 3rd edition, 1976.
Caras, Roger, *Going to the Zoo with Roger Caras*, 1973 (J), and *A Zoo in Your Room*, 1975 (J).
Carrington, Richard, *The Mammals*, 1967 (J).
Gergely, Tibor, *Animals: A Picture Book of Facts and Figures*, 1974 (J).
Grzimek, Bernhard, ed., *Grzimek's Animal Life Encyclopedia*, 1972-1975.
Johnson, Sylvia, *The Wildlife Atlas*, 1977 (J).
McClung, Robert M., *Lost Wild Worlds*, 1976.
National Geographic Society, *Vanishing Wildlife of North America*, 1974; *Wild Animals of North America*, 1979; *Wildlife Alert*, 1980 (J).
Wildlife Management Institute, *Big Game of North America*, 1978.

MONOTREMES:
Shuttlesworth, D., *The Wildlife of Australia and New Zealand*, 1967 (J).

MARSUPIALS:
Frith, H. J., and J. H. Calaby, *Kangaroos*, 1969.
Jenkins, Marie M., *Kangaroos, Opossums, and Other Marsupials*, 1975 (J).
Lauber, P., *The Surprising Kangaroos and Other Pouched Animals*, 1965 (J).
Lavine, Sigmund A., *Wonders of Marsupials*, 1978 (J).
Sherman, Geraldine, *Animals with Pouches: The Marsupials*, 1978 (J).

INSECTIVORES:
Griffiths, G. D., *The Story of a Hedgehog*, 1977 (J).
Ripper, Charles L., *Moles and Shrews*, 1957 (J).
Sheehan, Angela, *The Hedgehog*, 1977 (J).

CHIROPTERANS:
Lauber, P., *Bats: Wings in the Night*, 1968 (J).
Leen, Nina, *The Bat*, 1976 (J).
Yalden, D. W., and P. A. Morris, *The Lives of Bats*, 1976.

PRIMATES:
Allen, Martha D., *Meet the Monkeys*, 1979 (J).
Altmann, Stuart A., ed., *Social Communication Among Primates*, 1967.
Amon, Aline, *Orangutan: Endangered Ape*, 1977 (J), and *Reading, Writing, Chattering Chimps*, 1975 (J).
Eimerl, Sarel, and Irven DeVore, *The Primates*, 1965 (J).
Fenner, Carol, *Gorilla, Gorilla*, 1973 (J).
Hamburg, D. A., and E. R. McCown, eds., *The Great Apes*, 1979.
Jolly, Alison, *The Evolution of Primate Behavior*, 1972.
Lawick-Goodall, Jane van, *In the Shadow of Man*, 1971.
Leen, Nina, *Monkeys*, 1978 (J).
MacKinnon, John, *In Search of the Red Ape*, 1974.
Napier, Prue, *Chimpanzees*, 1976 (J).
Schaller, G., *The Mountain Gorilla*, 1963.
Teleki, G., L. Baldwin, and M. Rucks, *Aerial Apes: Gibbons of Asia*, 1979 (J).

EDENTATES:
Hopf, A. L., *Biography of an Armadillo*, 1975 (J).
Johnson, Sylvia A., *Animals of the Tropical Forests*, 1976 (J).

LAGOMORPHS:
Lockley, R. M., *The Private Life of the Rabbit*, 1974.
Orr, Robert T., *The Little-Known Pika*, 1977.
Silverstein, Alvin, *Rabbits: All About Them*, 1973 (J).

RODENTS:
Brady, Irene, *Beaver Year*, 1976 (J).
Hanney, Peter W., *Rodents: Their Lives and Habits*, 1975.
Newton, James R., *The March of the Lemmings*, 1976 (J).
Scott, Jack D., *Little Dogs of the Prairie*, 1977 (J).
Tunis, Edwin, *Chipmunks on the Doorstep*, 1971 (J).

CETACEANS, SIRENIANS, and PINNIPEDS:
Coerr, Eleanor, and William E. Evans, *Gigi: A Baby Whale Borrowed for Science and Returned to the Sea*, 1980 (J).
McClung, Robert M., *Hunted Mammals of the Sea*, 1978 (J).
McNulty, Faith, *The Great Whales*, 1974, and *Whales*, 1975 (J).
Scheffer, Victor B., *The Year of the Whale*, 1969.
Slijper, E. J., *Whales*, 2nd edition, 1979.
Time-Life Television, *Whales and Other Sea Mammals*, 1977 (J).

CARNIVORES:
Adamson, Joy, *Born Free* (lions), 1974 (J), *Living Free* (lions), 1961 (J), and *The Spotted Sphinx* (cheetahs), 1969.
Bueler, Lois E., *Wild Dogs of the World*, 1973.
Caputo, Robert, and Miriam Hsia, *Hyena Day*, 1978 (J).
Kruuk, Hans, *Hyaena*, 1975.
Leen, Nina, *Cats*, 1980 (J).
Maxwell, Gavin, *Ring of Bright Water* (otters), 1961.
Mech, L. David, *The Wolf*, 1970.
Patent, D., *Raccoons, Coatimundis, and Their Family*, 1979 (J).
Rue, Leonard Lee, III, *The World of the Red Fox*, 1969.
Ryden, Hope, *God's Dog* (coyotes), 1979, and *The Wild Pups: The True Story of a Coyote Family*, 1975 (J).
Schaller, G., *Golden Shadows, Flying Hooves* (lions), 1973, and with K. Schaller, *Wonders of Lions*, 1977 (J).
Silverstein, Alvin, *Cats: All About Them*, 1978 (J).
Steiner, Barbara A., *Biography of a Wolf*, 1973 (J).
Time-Life Television, *The Cats*, 1976 (J).
Wayre, Philip, *The River People* (otters), 1976.
Zim, Herbert S., *The Big Cats*, 1976 (J), and *Little Cats*, 1978 (J).

PROBOSCIDEANS:
Burger, Carl, *All About Elephants*, 1965 (J).
Douglas-Hamilton, Iain and Oria, *Among the Elephants*, 1975.
Hanks, John, *A Struggle for Survival: The Elephant Problem*, 1979.
Time-Life Television, *Elephants and Other Land Giants*, 1976 (J).

HYRACOIDS:
Kingdon, Jonathan, *East African Mammals*, 1971.

PERISSODACTYLS:
Arundel, Jocelyn, *Land of the Zebra*, 1974 (J).
Groves, Colin, *Horses, Asses and Zebras in the Wild*, 1974.
MacClintock, Dorcas, *A Natural History of Zebras*, 1976 (J).
Mochi, Ugo, and T. Donald Carter, *Hoofed Mammals of the World*, 1971.
Ryden, Hope, *Mustangs: A Return to the Wild*, 1978.
Willoughby, David P., *The Empire of Equus*, 1974.

ARTIODACTYLS:
Brown, Louise G., *Giraffes*, 1980 (J).
Geist, Valerius, *Mountain Sheep and Man in the Northern Wilds*, 1975.
Hopf, A. L., *Biography of a Giraffe*, 1978 (J), and *Pigs Wild and Tame*, 1979 (J).
Jenkins, Marie M., *Deer, Moose, Elk and Their Family*, 1979 (J).
Kilpatrick, Cathy, *Giraffes*, 1980 (J).
MacClintock, Dorcas, *A Natural History of Giraffes*, 1973 (J).
Matthiessen, Peter, *Oomingmak: The Expedition to the Musk Ox Island in the Bering Sea*, 1967.
Perry, Roger, *Wonders of Llamas*, 1977 (J).
Peterson, Randolph L., *North American Moose*, 1955.
Ricciuti, Edward R., *Wildlife of the Mountains*, 1979.
Schaller, G., *The Deer and the Tiger: A Study of Wildlife in India*, 1967, and *Stones of Silence: Journeys in the Himalaya*, 1980.
Schlein, Miriam, *Giraffe: The Silent Giant*, 1976 (J), and *On the Track of the Mystery Animal: The Story of the Discovery of the Okapi*, 1978 (J).
Stadtfeld, Curtis K., *Whitetail Deer: A Year's Cycle*, 1975.
Trevisick, C. H., *Hippos*, 1980 (J).
Van Wormer, Joe, *The World of the Pronghorn*, 1969.
Whitehead, G. Kenneth, *Deer of the World*, 1972.

The Orders of Mammals

Every mammal belongs to one of 19 scientific orders. The chart below lists the orders and presents some characteristics that distinguish their members. The chart also groups all the animal entries in the *Book of Mammals* by order.

Monotremes (*say* MON-uh-treemz)	A monotreme lays eggs. Each egg hatches into a young animal that laps milk from pores on its mother's belly: *echidna, platypus*
Marsupials (*say* mar-SOO-pea-ulz)	Marsupials usually have pouches. These animals give birth to tiny, underdeveloped young: *bandicoot, cuscus, kangaroo, koala, marsupial mouse, numbat, opossum, phalanger, quokka, sugar glider, Tasmanian devil, wallaby, wombat*
Insectivores (*say* in-SEK-tuh-vorz)	*Insectivore* means "insect eater." Most insectivores have long, narrow snouts and sharp claws that are well suited for digging for food, usually insects: *gymnure, hedgehog, mole, shrew, solenodon, tenrec*
Dermopterans (*say* der-MOP-tuh-runs)	A dermopteran glides from tree to tree. It does this by stretching well-developed folds of skin that extend from the sides of its neck to all four feet and that enclose its tail: *flying lemur*
Chiropterans (*say* kye-ROP-tuh-runs)	A chiropteran is a winged mammal, the only kind of mammal that actually flies: *bat*
Primates (*say* PRY-mates)	Primates have the ability to grasp with their fingers. They have hard nails on some of their fingers. Many primates can see objects in three dimensions. Some have highly developed brains: *aye-aye, bush baby, chimpanzee, gibbon, gorilla, lemur, loris, monkey, orangutan, potto, tarsier*
Edentates (*say* ee-DEN-tates)	*Edentate* means "without teeth," but all edentates except anteaters do have small teeth, usually in the back of the jaw: *anteater, armadillo, sloth*
Pholidotes (*say* FOLL-ih-dotes)	Large, horny scales cover the long, tapering body of a pholidote: *pangolin*
Lagomorphs (*say* LAG-uh-morfs)	Most lagomorphs have hind legs that are suited for leaping. Like a rodent, a lagomorph has chisel-like front teeth that grow throughout its life. But it has two pairs of front teeth in its upper jaw: *hare, pika, rabbit*
Rodents (*say* ROE-dents)	A rodent has chisel-like front teeth, one pair in the upper jaw and one pair in the lower jaw, that grow throughout the animal's life: *agouti, beaver, capybara, chinchilla, chipmunk, coypu, dormouse, gerbil, guinea pig, hamster, hutia, jerboa, kangaroo rat, lemming, mara, marmot, mole rat, mouse, muskrat, paca, pocket gopher, porcupine, prairie dog, rat, squirrel, tuco-tuco, vizcacha, vole*
Cetaceans (*say* see-TAY-shunz)	A cetacean spends its entire life in the water. It breathes air through a blowhole in the top of its head and feeds on fish, squid, or plankton: *porpoise, whale*
Carnivores (*say* CAR-nuh-vorz)	Most carnivores are meat eaters, but many also eat plants. They usually have sharp teeth that are well suited for cutting and tearing flesh. Some have claws that help them seize prey: *aardwolf, badger, bear, binturong, bobcat, cat, cheetah, civet, coati, coyote, dingo, dog, ferret, fossa, fox, genet, grison, hyena, jackal, jaguar, kinkajou, leopard, linsang, lion, lynx, marten, mink, mongoose, otter, panda, polecat, raccoon, raccoon dog, ratel, ringtail, skunk, suricate, tiger, weasel, wolf, wolverine, zorilla*
Pinnipeds (*say* PIN-ih-pedz)	*Pinniped* means "fin-footed," and a pinniped has four long flippers. These meat eaters spend much of their lives in water, but they come ashore to give birth: *seal and sea lion, walrus*
Tubulidentates (*say* too-byu-luh-DEN-tates)	This order was created just for the aardvark because of the tubelike teeth in the back of its jaw: *aardvark*
Proboscideans (*say* PROE-buh-SID-ee-unz)	A proboscidean has a long, flexible trunk and hooflike nails: *elephant*
Hyracoids (*say* HIGH-ruh-koidz)	A hyracoid has hooflike nails on its toes. Its upper front teeth are a little like tusks: *hyrax*
Sirenians (*say* sigh-REE-nee-unz)	A sirenian lives in shallow coastal waters and rivers where it feeds on water plants. It has paddlelike forelimbs and a flattened tail: *dugong, manatee*
Perissodactyls (*say* puh-RISS-uh-DAK-tulz)	A perissodactyl is a hoofed animal with an odd number of toes on each foot: *ass, horse, rhinoceros, tapir, zebra*
Artiodactyls (*say* art-ee-oh-DAK-tulz)	An artiodactyl is a hoofed animal with an even number of toes on each foot: *antelope, babirusa, bison, buffalo, camel, caribou, chamois, cow, deer, elk, gazelle, gerenuk, giraffe, goat, hippopotamus, hog, ibex, impala, llama, moose, musk-ox, okapi, peccary, pronghorn, serow, sheep, tahr, wildebeest, yak*

Index

Library of Congress CIP Data
Main entry under title:

National Geographic book of mammals.

Bibliography: p.
Includes index.
Summary: A picture encyclopedia presenting a general introduction to the world's mammals and the vital statistics and behavior for each of the entries from aardvark to zorilla.
1. Mammals—Dictionaries, Juvenile. [1. Mammals—Dictionaries]   I. Crump, Donald J.   II. Silcott, Philip B.   III. Smith, Ned.   IV. Sweet, Darrell K.
V. National Geographic Society (U. S.). Special Publications Division.   VI. Title: Book of mammals.
QL706.2.N37   599'.003'21   80-7825
AACR2

ISBN 0-87044-377-1 (v. 1: regular binding)
ISBN 0-87044-378-X (v. 2: regular binding)
ISBN 0-87044-376-3 (v. 1 and 2: regular binding)
ISBN 0-87044-380-1 (v. 1: library binding)
ISBN 0-87044-381-X (v. 2: library binding)
ISBN 0-87044-379-8 (v. 1 and 2: library binding)

Bold and majestic, a tiger stares from a forest in India.

Photographers' Credits

Composition for the *National Geographic Book of Mammals* by National Geographic's Photographic Services, Carl M. Shrader, Chief, Lawrence F. Ludwig, Assistant Chief. Printed and bound by Holladay-Tyler Printing Corp., Rockville, Md. Color separations by The Lanman Companies, Washington, D.C.; Progressive Color Corp., Rockville, Md.; Stevenson Photo Color Company, Cincinnati, Oh.

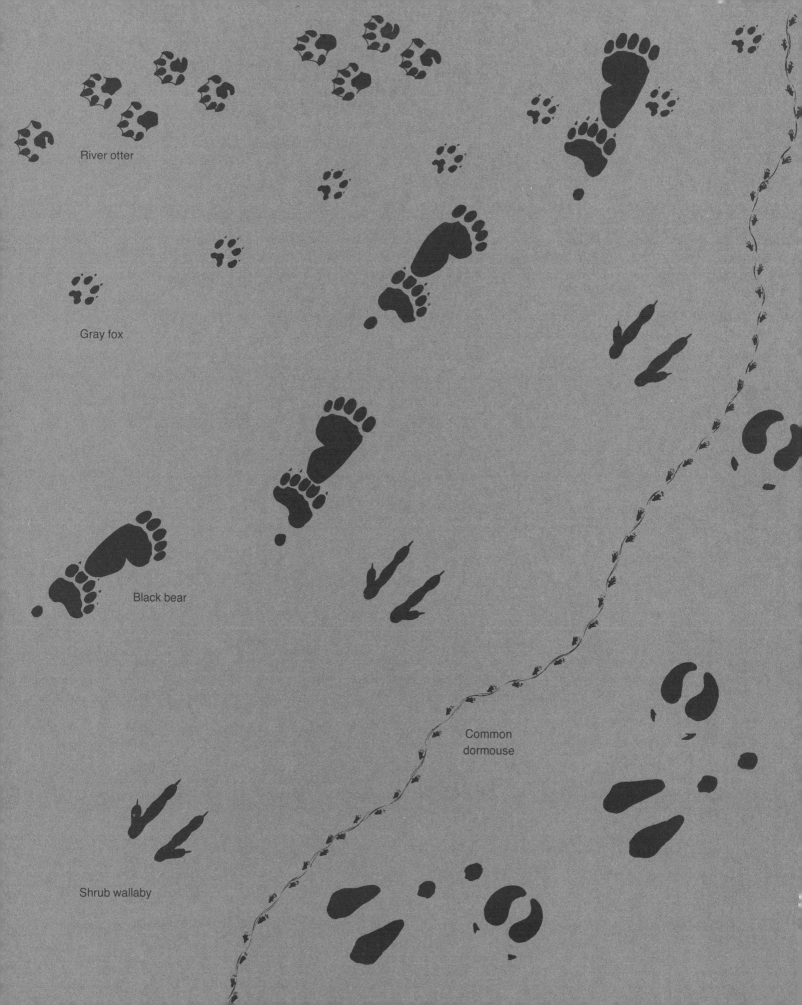

River otter

Gray fox

Black bear

Common
dormouse

Shrub wallaby